Ethics in Science and Engineering

Scrivener Publishing
3 Winter Street, Suite 3
Salem, MA 01970

Scrivener Publishing Collections Editors

James E. R. Couper	Ken Dragoon
Richard Erdlac	Rafiq Islam
Pradip Khaladkar	Vitthal Kulkarni
Norman Lieberman	Peter Martin
W. Kent Muhlbauer	Andrew Y. C. Nee
S. A. Sherif	James G. Speight

Publishers at Scrivener
Martin Scrivener (martin@scrivenerpublishing.com)
Phillip Carmical (pcarmical@scrivenerpublishing.com)

Ethics in Science and Engineering

James G. Speight
and Russell Foote

Copyright © 2011 by Scrivener Publishing LLC. All rights reserved.

Co-published by John Wiley & Sons, Inc. Hoboken, New Jersey, and Scrivener Publishing LLC, Salem, Massachusetts.
Published simultaneously in Canada.

No part of this publication may be reproduced, stored in a retrieval system, or transmitted in any form or by any means, electronic, mechanical, photocopying, recording, scanning, or otherwise, except as permitted under Section 107 or 108 of the 1976 United States Copyright Act, without either the prior written permission of the Publisher, or authorization through payment of the appropriate per-copy fee to the Copyright Clearance Center, Inc., 222 Rosewood Drive, Danvers, MA 01923, (978) 750-8400, fax (978) 750-4470, or on the web at www.copyright.com. Requests to the Publisher for permission should be addressed to the Permissions Department, John Wiley & Sons, Inc., 111 River Street, Hoboken, NJ 07030, (201) 748-6011, fax (201) 748-6008, or online at http://www.wiley.com/go/permission.

Limit of Liability/Disclaimer of Warranty: While the publisher and author have used their best efforts in preparing this book, they make no representations or warranties with respect to the accuracy or completeness of the contents of this book and specifically disclaim any implied warranties of merchantability or fitness for a particular purpose. No warranty may be created or extended by sales representatives or written sales materials. The advice and strategies contained herein may not be suitable for your situation. You should consult with a professional where appropriate. Neither the publisher nor author shall be liable for any loss of profit or any other commercial damages, including but not limited to special, incidental, consequential, or other damages.

For general information on our other products and services or for technical support, please contact our Customer Care Department within the United States at (800) 762-2974, outside the United States at (317) 572-3993 or fax (317) 572-4002.

Wiley also publishes its books in a variety of electronic formats. Some content that appears in print may not be available in electronic formats. For more information about Wiley products, visit our web site at www.wiley.com.

For more information about Scrivener products please visit www.scrivenerpublishing.com.

Cover design by Russell Richardson.

Library of Congress Cataloging-in-Publication Data:

ISBN 978-0-470-62602-3

Contents

Preface

The various scientific and engineering disciplines are world-wide professional disciplines. The members of these disciplines collect factual data, and the ensuing treatment of the data to discover new arenas of knowledge is universal. No one can foresee the tortuous path of scientific and engineering investigation and know where experimentation and observation may lead. Then there is always the mode of data interpretation.

The pursuit of science and engineering requires freedom of thought and, in the academic sense, unrestricted communication. It is through the professionalism of the members of the scientific and engineering disciplines that world knowledge and technology advances. Yet there are continuous reports of unethical behavior in the form of data manipulation, cheating, and plagiarism at the highest levels of the disciplines. The causes are manifold whether it is the need to advance in one of the chosen disciplines or to compete successfully for and obtain research funding.

Disappointingly, individuals who are oriented to any form of scientific or engineering dishonesty are individuals who had previously displayed little or no consideration for the feelings of others and are therefore more interested in other scientists and engineers recognizing them by any means necessary.

This project was triggered by a combination of factors – awareness by the authors of the increasing frequency of unethical practices in the realm education, recognition of the focus of the literature on ethical behavior, and the realization that ethical/unethical behavior is the outcome of choice and is not due to chance. There is no attempt to be judgmental but to encourage everyone to reflect on themselves philosophically (that is, in terms of individual values and beliefs) since it was absolutely clear that personal motivations and preferences can override any other contributory factor.

This book gives an account of various scientific and engineering disciplines and examines the potential for unethical behavior by professionals. Documented examples are presented to show where the matter could have been halted before it became an ethical issue. The authors also look to the future to see what is in store for professionals in the scientific and engineering disciplines and how the potential for unethical behavior can be negated.

On the basis of the observations and research by the authors, this publication seeks to advance basic requirements for the application of ethical behavior, to mitigate the frequent occurrences of misconduct, which currently and frequently appear in the scientific and engineering disciplines.

To claim completeness in a project of this nature would be foolhardy, but it is hoped that this preliminary treatment will stimulate discussions about ethics among students and faculty within universities and other educational institutions. There is a further hope that such self-examination will encourage students and faculty to raise their own standards of ethical conduct without having to be forced to do so.

Russell Foote, PhD
Trinidad and Tobago

James G. Speight, PhD, DSc
Laramie Wyoming, USA

1

Explaining Ethics

1.1 Introduction

Scientific and engineering disciplines are considered to be highly ethical professions in which scientists and engineers exhibit behavior of the highest ethical and moral standards.

Ethics is "the normative science of conduct, and conduct is a collective name for voluntary actions" (Lillie, 2001, page 3). In this regard, voluntary actions are those actions that could have been done differently, where such actions may be good or bad, right or wrong, or moral or immoral. Ethics focuses not on what people think but what they ought to think or do. An ethical science is an in-depth, systematic study of the standards for judging right and wrong, good and bad principles, guiding means, and how far we will or should go (Lillie, 2001; Howard and Korver, 2008).

Generally, ethics (morality) is a core branch of philosophy that attempts to define right and wrong; what a scientist or engineer *ought* to do is as distinct from what they may do. In philosophical studies, ethics is usually divided into three sub-fields: (1) meta-ethics, (2) normative ethics, and (3) applied ethics.

Meta-ethics includes investigation of whether or not ethical claims are capable of being true or false, or if they are expressions of emotion. *Normative ethics* attempts to arrive at practical moral standards that would tell, for example, the scientist or engineer what is right or what is wrong. *Applied ethics* is the application of theories of right and wrong and theories of value to specific issues such as honesty and lying.

Whatever the definition, ethics is one of the pillars of scientific research, teaching and community service requirements of higher education. It is definitely one of the criteria for evaluating the quality of higher education in these aforementioned areas. Despite the variables that contribute to ethical or unethical behavior, the central determinants are the personal thoughts and behavior of the scientist and engineer which determines the meaning that an individual assigns to their position regarding ethics.

Personal thoughts and behavior can override the influence of any other factor, including the Codes of Ethics of professional bodies. The ability to manage emotions during the processes of scientific and engineering research orients many individuals to act on feelings and engage in unethical practices. This is reflected in the increasing frequency of reports of misconduct in the scientific and engineering disciplines (Chapter 8).

The realm of ethics is concerned with standards and requirements for socially acceptable behavior, in addition to following proper procedures for getting things done at any level of interaction – individual, group, organizational, community, governmental or regional. Ethics has several strands that are applicable to the scientist and engineer: (1) *descriptive ethics*, which the actual behavior of people and the ethical requirements of their behavior, (2) *normative ethics*, which is the application of the values that are good enough to guide interaction, and (3) *applied ethics*, which is the application of normative rights to specific issues, disciplines and settings (Kitchener and Kitchener, 2009, page 5–6).

Furthermore, the ethical aspects of scientific and engineering research revolve around the proper method to collect, analyze and report all aspects of a study, and the responses to researcher-respondent interactions, which are especially true in the social sciences where surveys of human actions are accumulated (Kitchener and Kitchener, 2009, page 6).

The requirements, in this regard, are stipulated in various Codes of Ethics documents of scientific and engineering organizations such as: The American Chemical Society (ACS), The Royal Society of Chemistry (RSC), The American Institute for Chemical Engineers (AIChE), The Society of Petroleum Engineers (SPE), The American Psychological Association (APA), the American Sociological Association (ASA), American Anthropological Association (AAA) and various other disciplinary bodies across the world.

However, such codes (Chapter 6) do not resolve the issue which, in the final analysis, depends on personal decision-making, and freedom from bias, prejudice and personal values (Kitchener and Kitchener, 2009, page 32). Furthermore, these codes cannot, and must not, be ignored by using claims of academic freedom. Generally, they are intended to legally reinforce the need for respect for all other human beings independent of what anybody thinks about location, upbringing, gender, ethnicity, religious affiliation, age, culture, level of education and other characteristics.

In fact, academic integrity is critical to higher education, especially where research and learning manifest. However, the incidence of academic dishonesty in university settings leaves much to be desired with occurrences of dishonesty among 40% to 70% of the students (Davis et al., Kibler, 1998; Marcoux, 2002). However, faculty consensus is limited on what forms of behavior constitute dishonesty. Traditional forms of academic dishonesty, where there is consensus, such as looking on another student's paper during a test or handing in work done by a classmate, have changed with technological advances (Marcoux, 2002). Modern computer programs and applications, Internet access to diverse and instant information, distance learning classes, and hand-held computing devices which can transmit information in moments change the need for an increasing awareness by faculty, in terms of addressing academic dishonesty.

Indeed, ethical issues have come and will remain at the fore because of the prioritization of differences by scientists and engineers as they seek to attain a more privileged position in their organization and the world of academia. This behavior has been compounded further by the emergence of procedural inconsistencies in several major research projects (Kitchener and Kitchener, 2009, page 8) (see also Chapter 8).

In addition, there seems to be much truth in the post-modern view of research ethics that every research activity, question and decision has ethical underpinnings. In such cases, a number of pertinent and revealing questions should follow with the corresponding ethical issues identified.

Moreover, honesty has to be practiced at all times and must be evaluated on the basis of intentions and not outcomes, like some occupations. However, "intentions will stop being regarded as good if they repeatedly produce bad results or no results at all" (Lillie, 2001, page 13). In addition, the correctness of an action depends on the action as a whole, not on past actions.

Whether a scientist or engineer's conduct is good or bad may be: (1) instinctive and discernible through one's actions, (2) intentional, which may be direct and motivating, or (3) indirect, rooted in desire, which is a consciousness to act in a particular manner, or (4) a matter of calculated choice (Lillie, 2001, page 24–33).

Furthermore, explanations of theories of ethical behavior have been described as: (1) absolute, which assumes that changes in circumstances make no difference in the rightness or wrongness of guidelines for action, (2) relative, which indicates that ethical conduct can vary from person to person, (3) naturalistic, which is due to the variation of ethical standards with a person's attitude, in which case it is subjective, or, if ethical standards vary with a person's attitude changes, it is objective, (4) deontological, which is when correctness depends on the action itself and, (5) teleological, which focuses on correctness of actions in terms of levels of the benefits that result (Lillie, 2001, page 98–101).

Indeed, the actions of one person can impact on the actions of others and, as such, the general nature and direction of actions in a society may affect the choices of others and their level of consideration for moral standards. Such actions impact concerns for the common good, levels of egoism and altruism, and the eventual emergence of rights, duties and entitlements.

Ethical disagreements on rights, duties and entitlements are also possible and may take the form of disagreement in belief. This is when an individual believes in one aspect of a theory or argument, and another individual believes in a different aspect of the theory or argument such that one individual persistently challenges the other. Ethical disagreements may also take the form of disagreement in attitude. This is when an individual has a favorable, or unfavorable, attitude towards one side of the theory, and another individual has the opposite attitude towards one side of the theory (Stevenson, 2006). Such disagreements are typical of the types of disagreements which occur between scientist and engineers. But what really matters is the means by which a scientist or engineer reaches their conclusion, how the data were handled, and any ensuing interpretation of the data.

The extent and frequency of agreements and disagreements vary with the extent to which there exists an ethical environment, defined as "the climate of values in which people live and in which young people grow up" (Haydon 2006, page 2). Schools, like all other organizations, share an ethical environment. All societies have norms of conduct – norms are synonymous for morals which signify how people should treat each other. *Norm conformity* is recognized as an obligation or duty, in the absence of norms being identified, where people can be guided by the consequences of their actions.

Values, laws and religious teachings are part of the ethical environment (Haydon, 2006, page 35 and 37). As such, values and laws must be considered to evaluate the ethical environment, which may have to be changed, if necessary. This can happen through individual action, legal changes, or education. Implicit in the creation and maintenance of an ethical environment is the emergence of regimes of reason and unreason, which are comprised of conscious and unconscious, opposing and accepted values that often clash with each other in a society (Leitch, 1992, page 1–3).

The assessment of rights, duties, and entitlements is also a moral issue. Moral capacities and judgments would have been shaped by personality, socialization, situational demographic (age, gender, ethnicity etc.) and broader societal factors. Generally, scientists and engineers act because they want to achieve a goal by which they satisfy an interest or desire (Furrow, 2005, 10). These factors do not act independently of each other, rather in combination. Indeed, morally appropriate behavior is driven by thoughts and feelings that were cultivated and reinforced across time and space.

Moral autonomy is not achievable when personal desires, emotions and inclinations persistently influence a person's judgment. However, moral autonomy has to be exercised within certain societal boundaries even if it conflicts with individual's needs. In this regard, it is necessary to evaluate desires and goals (Furrow 2005, page 25).

It follows that reasoning is instrumental in helping scientists and engineers pursue and attain certain goals. However, caution is warranted because reasoning may be either rational or emotionally loaded. In fact, the reality of cultural differences – individual, group, and organizational has universally generated a diversity of moral codes where people do not subscribe to a single moral code. This has resulted in "relative morality," which does not mean that there is no true objective moral code. Relative morality has been justified on the basis of physical and cultural differences, and the constant promotion of tolerance for different views (Rachels, 2000, page 12). In the context of social changes, communication and interactions with other countries, there has been significant cross-fertilization of ideas influencing people to make judgments on levels of morality (Furrow 2005, page 38).

It is generally known that once a promise or commitment is made there is an obligation to keep it. Some scientists and engineers may not keep their obligations because they are not quite comfortable with themselves, or because of others giving them differing advice. The result is diminished willpower, or intention, to fulfill their obligation. Intentions are the outcomes of deliberating with self to decide what to do (Williams, 2006, page 18).

While beliefs are not always under voluntary control, it is true that there is a choice of what to believe, and, as a result, choice is controlled. In this regard, the scientist and engineer must remain open-minded and always be ready to evaluate arguments and findings from different perspectives.

Consequently, it must be recognized that the end does not justify the means, a rational basis must be established for dealing with uncertainty in any type of research, some types of research may not be ethically justifiable, and, while researchers prefer to minimize errors, there are those who prefer false positives over false negatives (Shrader-Frechette, 1994).

If the act that the individual scientist or engineer performs is in their power not to perform, then they are responsible for that act and must face the consequences (Chisholm, 2008, page 418). This would establish the morality of the action. It must be noted, however, that the orientation to autonomous or independent individual-level action is shaped and reshaped by a changing society. As a result, the central influencing factor is the quality of individual-level socialization despite the changing nature of the context. It is further reinforced by law enforcement, cultural influences, accountability arrangements, and monitoring and evaluation standards. In addition to these, the promotion of equity initiatives (Kezar et al., 2008, page 154–56) would serve to reduce ethical lapses in universities and other settings.

1.2 The Impact of Science and Engineering

Scientific and engineering are the driving forces for the majority of changes witnessed in the 20th century. They require a critical mind that is free of prejudice and open to new ways of thinking, with the capability to apply honest principles by investigators. The rapid development of modern science and engineering since the Renaissance is due mainly to the postulate that scientific theories should be independent of theological or religious beliefs. In the 17th and 18th century, knowledge was mainly exchanged through scientific academies which disseminated new theories and thus accelerated scientific progress. At the beginning of the 19th century, there was a remarkable rise in academic research at universities (*pure research* and *basic research*) and many university-based scientists were not interested in the technological applications of the results of their endeavors.

On the other hand, even though the research methods in industry (*industrial research* and *applied research*) differed and emanated from basic research, each method had completely different aims and rules. The focus was to acquire new knowledge *and* to adapt this knowledge to produce a profitable product for sale. The results were not the property of the investigating scientists and engineers, but the property of the industry for which they worked.

Generally, discussions concerning ethical problems were more or less absent from both realms. In academia, scientists and engineers were indifferent to the possible consequences of their work, and in industry, employers did not consider it appropriate for scientists and engineers to worry about ethical problems. In fact, this atmosphere may still exist in many scientific and engineering laboratories.

At the beginning of the 21st Century, changes in the interactions between scientist and engineers from different universities are taking place and scientists and engineers in academia and industry are increasingly collaborating. Furthermore, the results of industrial research are often published in peer-reviewed journals. As a result, it has become pertinent and necessary to evaluate, from an ethical point of view, not only the use of scientific and engineering knowledge, but also its production (Iaccarino, 2001).

As a result of the knowledge explosion, the impact of science and engineering is reflected in several ways. For example, the focus on science and engineering recognizes that new fundamental knowledge and technology will lead to the creation of new industries with associated high technology. In addition, a clean energy future, through expanded investment in research, development, demonstration, and deployment of clean energy technologies, can help reduce dependence on domestic and imported oil. It also can create green jobs, and limit the impact of climate change. The development of better science and technology is improving the prediction and prevention of, and the reaction to, destabilizing or paralyzing natural and man-made threats; improving capabilities for bio-defense; and monitoring nuclear nonproliferation compliance and preventing the surreptitious entry of weapons of mass destruction (NSB, 2010).

In addition, public attitudes about emerging areas of scientific and engineering research and new technologies will have an impact on innovation. The climate of opinion concerning new research areas could influence levels of public and private investment in related technological innovations and, eventually, the adoption of new technologies and the growth of industries based on these technologies. Furthermore, public opinion is swayed by the occurrence of cheating and misconduct in science and engineering.

On the issue of cheating and misconduct, students use a variety of methods to cheat on class examinations (Bernardi et al., 2008). In order to preserve the integrity of science and engineering, teachers and professors must: (1) acknowledge that cheating occurs, (2) examine the level of cheating, and (3) determine the reasons for cheating. Then actions such as having multiple versions of the examination and scrambling the questions on these versions would be a start to deter cheating. In addition, punishment of these actions through expulsion from the program or another equally drastic measure will force the students to recognize that there is no tolerance for cheating and misconduct. Such actions are necessary for science and engineering to remain honorable disciplines. This will preserve the beneficial impacts of the scientific and engineering disciplines.

Furthermore, there are other critical issues relating to the assessment of impact of the work of scientific and engineering professionals: (1) the impact of any new technology or modified technology takes time, and (2) the measure of the impact is not achieved by the use of a so-called standard citation index, which is used to indicate the importance of journals and the papers contained therein.

There is always the distinct possibility that the number of citations is directly related to those who are critics of the work and may consider it nothing short of ludicrous (Did the reviewers concentrate on grammatical errors rather than on scientific content?). Not all papers in high quality publications are of great significance, and high quality papers can appear in lower quality publication media. Therefore, the academic form of evaluation can be severely underwhelming and even incorrect!

On the other hand, the young professional's supervisor may fail to recognize the impact of the work, especially if the name of the young professional name is not included as a co-author. The rationale for such an omission is not easy to explain and must often remain in the dark recesses of the mind of the supervisor.

If the scientist or engineer request an evaluation of their work and its effects, evaluators should be selected from academic or company colleagues, and even users – if the concept has been reduced to practice.

Some academic institutions and companies prohibit such methods of evaluation from writers not having an academic affiliation or a company affiliation, respectively. This can be a serious blow to the morale of scientists and engineers because some of the field's best researchers work at other institutions.

Unless such an assessment of the work can be performed, the young professional may fail because the significance of their work may be ignored. It is probable that the young professional believes the impact of his work is not recognized, therefore, bypassing new ideas and techniques. The world is visibly marked by science and engineering.

The scientific and engineering disciplines continue to move towards new and important discoveries which continue to have crucial consequences for society. As a result, scientists, engineers, and the public in general, should be concerned about the consequences of the correct or incorrect data that drive these discoveries.

As a whole, this scientific revolution generates a new system of values and creates conditions which must involve an *ethical approach*. In managing new discoveries, scientists and engineers are faced with economic competition, which is combined with ideology and serves as a basis for scientific effort. This highlights the responsibility of scientists and engineers and calls for them to reaffirm a generation of older values and then create a set of new ethical values.

There is a current challenge to develop workable frameworks by which ethics and ethical behavior can be defined and the concepts followed. It is hoped that by doing so, the cheating and misconduct (which seems to propagate as the years pass) can be diminished with science and engineering, affording a positive influence on the future.

1.3 The Framework of Ethics

Ethics is based on feelings and instinct, which provides information that allows ethical choices to be made. In addition, ethics does not necessarily involve following cultural law. Some cultures may be ethical while other cultures are corrupt or ignore ethical concerns – following the old adage, *when in Rome, do as the Romans do*, is not a satisfactory ethical standard. On the other hand, ethics provides many reasons for how scientist and engineers ought to act (Markkula, 2010).

One of the hurdles of applying ethics to science and engineering is to find the correct place to start. For example, one of the most vital areas of modern philosophical debate concerns the hands-on practice of science and engineering and the treatment of the data. If a scientist or engineer begins with the premise that their actions are always moral, this reflects their attitude to helping humanity in general. They may conclude that their actions were correct and what was written on paper was infallible, and therefore, the reason for the ten additional experiments used to produce a possible answer to the problem.

Such attitudes are, in fact, the starting point of much of the traditional moral philosophy, as applied to, science and engineering. It is at the heart of the distinction between what is right and what is wrong with many scientists and engineers. The scientist and engineer had burned the late candle in bringing his model to a conclusion, but has forgotten that many of his assumptions are invalid. Similarly, the he has toiled in the laboratory to complete the additional experiments that were invariably designed to prove his theory without even acknowledging that the theory could be irrational.

Part of the difficulty with applied ethics is exemplified by the very real issue of the relationship between facts and values in the matter of under study.

It is clear, from further and more detailed consideration of the issues above (The model cannot be wrong or the experiments were designed impartially.) that:

1. scientific and engineering ethics (morality) require a *human agent* (the scientist or engineer) to carry out the actions and often, but not always, also a human as the recipient of the action,
2. the moral action requires the capacity within the scientist or engineer to reason with the actions, and then understand whether such actions are ethical or unethical (moral or immoral), and
3. the scientist or engineer must be responsible for his actions and have the freedom – in some cases it is designated as *academic freedom* (Chapter 5) – to act otherwise.

In addition, the role of reasoning, care, personality attributes, interactions, and the motivations of the individual have been variously emphasized. The scientist or engineer can act contrary to his or her embodied nature and according to a rational principle that transcends that nature (McLaren 2006, page 143). In fact, prospects for the theoretical synthesis of the contemporary perspectives of care ethics, cognitive developmentalism and character education are both good and bad insofar as: (1) the arguments of only two of these views converge, which is bad, and (2) morality has cognitive and emotional dimensions.

A much more empirically relevant position on ethics should consider human nature and standards of morality. However, in this era of social norm deterioration and moral confusion, individual beliefs and values cannot depend on social influences for a sense of direction and therein lies the tension to maintain certain ethical standards.

As a professional, if the moral reasoning of a scientist or engineer is inadequate, he should turn to The Codes of Ethics (Chapter 6) but these may not cover certain issues. The ultimate requirement of a professional is to benefit others, doing no harm, while being fair and faithful. (Welfel and Kitchener, 1999, page 134).

Research ethics however, seeks to ensure that scientific and engineering research is conducted within acceptable standards of morality, in order to preserve integrity, validity, and reliability of the study. While standards for conducting research focuses on the study itself, ethical issues emphasize people. Such issues include concerns about fraud, misconduct, harm to subjects, infringement of rights, manipulating the data directly or through the misuse of statistics, conflicts of interest, as well as misrepresentation of self and others. In fact, many professional bodies have stipulated codes of conduct to guide scientific practices.

Several theories of theories of ethics affirm that harm can emerge after a study is done and such harm is to be weighed. For example, *utilitarianism ethics* evaluates morality in terms of right and wrong while *deontological ethics* believes that some actions are inherently right or wrong despite consequences (Peach, 1995, page 15–17). While and focusing on the details of moral cases as well as providing procedures for resolution. *Virtue ethics* highlight human characteristics, habits, skills, traits, motivations, and intentions (Annas, 2006, page 516–517).

In the Protestant Ethic and the Spirit of Capitalism, a relationship between morality and work was established and a morality of aspiration focuses on rewards for outstanding performance rather than on punishment administered for failure.

In keeping with many non-scientists and non-engineers, some scientist and engineers (believe it or not) may not have a strong commitment to the process of rational thought, resulting in a focus on image (Kearney, 1999, page 12). Such occurrences render it possible for any mechanical expression of responsibility to be eroded when the new emerges. Socialization into responsibilities during childhood and teenage years is one of the prerequisites for ethical commitment in later years and also for the exercise of professionalism. One therefore has to be responsible first before one can become or act like a professional. The demonstration of responsibility cannot be talked into being. Where there are interactional bonds, there is a commitment to be responsible for the other and the resulting emergence of a sense of culture.

As responsibility develops, the stages involve inclinations to punish or obey, orientation to seek pleasure and avoid pain, the emergence of social awareness, and acceptance of the importance of ethics (Alcorn 2001, 86–88).

A framework has been proposed to analyze ethical issues in the behavioral sciences (Beauchamp et al., 1982, page 46) in which *harm* is defined as any situation where an individual's well-being is reduced, while a *benefit* occurs when well-being is enhanced. Harm, like benefit, can impact participants in scientific and engineering research as a result of the research or during the conduct of research. Such ethical issues are an invasion of privacy, loss of confidentiality, lack of informed consent or deception. These may take the form of stress or humiliation, affecting group interests and violating the norms of healthy interpersonal relationships (Beauchamp et al., 1982, page 104–109).

Thus there is a strong need for the scientist and engineer to clarify his perception of the moral issue, list alternative causes of action, make a choice from the options available, decide on the consequences while evaluating his values, and discuss with others in an attempt to gain further guidance in decision making (Smith, 1990, 146).

Ethics and morality are similar, but yet, different. Morality involves "sensitivity to the needs of others... and responsibility for taking care" (Walker, 2003, page 59). Furthermore, morality is reflected in the fair treatment of other people and the monitoring of relationships with others relative to the nature of the attachment. The particular context can frame and guide ongoing moral thinking and action, as well as making judgments and taking responsibility (Walker, 2003, page xii). Such contexts include but are not limited to classrooms, conferences, professional associations, order of author listing in publications, and graduate programs. Moreover thinking and theorizing about moral issues – valuing, judgment and responsibility – without paying attention to context is questionable, if not futile (Walker, 2003, page xiii).

Criticisms of theories on moral development (Gilligan, 1982) have led to the development of the more practical-oriented theories of caring and justice (to a lesser extent), which complement each other. Following from this, it is possible to derive ethical frames that foreground sharing, sensitivity to others, personal responsibility, and shared decision-making (Lincoln, 2009, page 157). In addition, *justice theories* emphasize legal procedures, natural rights, and individuality; whereas *care theories* can be criticized for ignoring the concerns of minorities and other marginalized groups (Lincoln, 2009, page 157).

Justice theories, like care theories, are applicable to community work in order to restore the balance of justice for issues of gender, social class, race, politics, history, poverty or oppression (Lincoln, 2009, page 157). The point of departure is the most significant contribution to our understanding of morality that links moral reasoning and behavior (Walker, 2004, page 2). This is involves focusing on the importance of moral values to one's identity, the sense of personal responsibility, which orients individuals to ensure that their actions are consistent with their moral judgments, and self-consistency, which emphasizes the need to maintain congruence between one's sense of morality and moral decision-making. (Walker, 2004, page 2–4).

On the other hand, it has been proposed that there are four dimensions of moral behavior: interpreting how the actions of others are affected by oneself, determining the ideal moral behavior for a situation, deciding which moral action to pursue, and acting on the decision (Bergman, 2004, page 25).

Identity theory postulates that individuals act on the basis of beliefs and values that collectively contribute to one's sense of self and moral identity (Moshman, 2004, 92). This is both discovery and creation as the individual scientist or engineer can decide what type of person they want to be (Moshman, 2004, 91). If there is not any judgment-action correspondence, then this is a false moral identity due to a weak moral commitment. Indeed, moral reflection is a capacity available in different forms at all points in development. People can be used to evaluate social or technical situations from a moral point of view (Nucci, 2004, page 127).

Feminism and racism have historically, and successfully, identified a major ethical void in the scientific and engineering disciplines by exposing practices that have denied access to women or other oppressed groups by ignoring or devaluing them (Brabeck and Brabeck, 2009, page 39 and page 41). To combat this prevailing attitude toward women and other races, it is necessary to raise awareness of depressing social conditions, promote key values of fairness, welfare, and justice, encourage rethinking of the social issues, and examine the historical aspects of domination, control, alienation and inequities (Thomas, 2009, page 54).

Another perspective that is applicable to ethical or unethical decision-making and actions is the *Theory of Planned Behavior*, which emphasizes that intentions influence volitional behavior by exerting a motivational effect on individuals (Kiriakidis, 2008, page 2211). In addition, the predictors of intentions are attitudes, subjective or personal norms, perceived behavioral control and hypothesized relationships; all of which have gained much empirical support (Kraft et al., 2005, page 480; Kiriakidis, 2008, page 2211).

Of significance are the methodological and empirical aspects of this issue since these interact and collectively influence the extent of the ethical commitments of a scientist or engineer in any context – academic or non-academic. As such it is not only an intellectual or purely rational exercise at the individual, group, organizational or societal levels – but it is a subjective exercise. Indeed, this is an issue where the natural emotionality of scientists and engineers provide a catalyst for the realization of ethical imperatives, requiring philosophical attention (Tangney et al., 2007, 345).

Then, the framework for ethical decision making lies in the following concepts (Markkula, 2010):

1. Recognize an Ethical Issue – is the decision or situation damaging to someone or to some group? Does the decision involve a choice between a good and bad alternative? Is the issue about more than what is legal or what is most efficient?

2. Investigate the Facts – what are the relevant facts of the issue? What individuals and groups have an important stake in the outcome? Have all the relevant persons and groups been consulted?
3. Evaluate Alternative Actions – especially the option which best respects the rights of all who have a stake and treats people equally or proportionately? Which option best addresses the situation?
4. Act and Reflect on the Outcome – how will the decision be implemented with attention to the concerns of all stakeholders? How will the decision turn out?

Whilst considering this framework, it is also worth considering the *The 4-Way Test* which is the *credo* or operating principle of Rotary International (Chapter 2).

1.4 Ethics in Professional Life

Scientists and engineers have become increasingly interested in questions of ethics. It might be diversity of the sub-disciplines or the fundamental questions from scientists and engineers which leads to this interest.

Some scientists and engineers are more enamored with such interests and discussions than others. However, the disciplines largely belie the common interpretation by non-scientists and non-engineers as more of a repository of descriptive facts about the world than some deeper intellectual perspective on their meaning.

To many scientists and engineers ethics is often misunderstood and believed to be, or seen as, an abstract and speculative area. An area that is as impractical as it is incomprehensible and is of interest only to scholars paid to think thoughts bearing little connection to reality outside the ivory tower.

However, it is vital that ethics not be treated as something remote to be studied only by scholars locked away in universities. Ethics deals with values, good and bad, and right and wrong. Scientists or engineers cannot avoid not being involved in ethics, for what they do and what they do not do is always subject to ethical evaluation.

Ethics, also known as moral philosophy, involves systematic intellectual reflection on morality in general – where morality is the realm of significant normative concerns, often described by thoughts of right or wrong, or specific moral concerns.

One realm of applied ethics that has received considerable attention in the scientific and engineering communities focuses on professional conduct. The moral questions asked by scientists and engineers, as well as those in, for instance, the fields of law, medicine or business, are legitimate components of ethical enquiry.

In addition to ethics involving both theoretical and applied concerns, another useful distinction can be drawn between descriptive ethics, normative ethics, and meta-ethics (though only the latter two are represented in philosophical literature). The aim of descriptive ethics is to characterize existing moral schemes; this has been an important feature of the science and engineering disciplines.

Normative ethics is devoted to constructing a suitable moral basis for informing human conduct, while meta-ethics, is more an examination of the characteristics of ethical reasoning, or systems of ethics.

Thus, in science and engineering, ethics typically involves reflection upon moral questions that arise in research, publication and other professional activities. From this several questions arise:

- Is it wrong to bend data to support one's conclusions?

- Is it wrong to publish data gathered under some assumption of confidentiality on the part of the research subject?
- Is it wrong to publish a work based substantially on the research of the graduate student(s) of a professor (mentor) as the professor's own?
- Is it wrong to enter the policy arena as a scientist or engineer, where objectivity and partiality could well clash?

The sheer number and complexity of these kinds of ethical issues, in the conduct of science, is amply evidenced in the older literature (Shrader-Frechette, 1994; Weeks and Kinser, 1994) and are addressed in this book (Chapter 8 and Chapter 9). However, this prevailing sense of ethics among scientists and engineers avoids, at least, as many difficult moral questions as are asked. Indeed, the kinds of moral issues entertained in much of scientific and engineering work tends to dodge (or ignore) the much larger moral question about the ethics of science and engineering. Also, whether the current research priorities and the amount of scarce resources that are currently allocated to scientific and engineering research are justifiable, and whether the typically elevated status of scientific and engineering pronouncements on reality is justified, in the light of the many stinging critiques of misconduct in science and engineering (Chapter 9).

One way to weave together both professional and substantive ethical behavior in science and engineering is to recall the heritage of both disciplines. The net result could be a greater affinity between professional and substantive concerns among scientist and engineers.

Professional ethics represents the context, or the process, out of which the content, or the result, of substantive ethics emerges. Joining these two areas will allow scientist and engineers to be properly reflexive in the moral statements they make about their work.

One of the most familiar areas of ethical enquiry in science and engineering involves research and analytical techniques. The act of research itself, and the consideration of the role of the researcher vis-à-vis the research subject(s), has also been a popular subject of enquiry. Another area of concern is related to the manner in which science and engineering are represented, and the direct social significance of the data. Indeed, ethical issues become more focused as one moves from a particular scientific or engineering concept to its technical implementation, and finally to its application.

There are so many questions which scientists and engineers could apply their intellectual efforts and organize these questions as presented below:

- What is the role of ethics in scientific and engineering and practice?
- What kinds of values have implicitly or explicitly accompanied the practice of science and engineering in recent history?
- Is it appropriate for only a subgroup of scientist and engineers to be intellectually concerned with ethics, or does ethics pertain to all scientists and engineers?
- How might scientists and engineers proceed to address ethical problems in their work?
- To what extent is ethical conduct desirable, definable and/or enforceable in the practice of science and engineering?

Finally, understanding ethics as an inextricable part of the work of scientists and engineers is the first step in their ability to answer these questions.

Going beyond much of the previous literature on ethics, the authors of this book attempt to show how ethical issues, despite its varied philosophical moorings, ultimately find fuel in and can be put in proper perspective on the basis of understanding human nature or – more correctly – the goals of scientists and engineers.

The goal now is to scientists and engineers more sensitive to the ethical implications of their work. This requires a start from the basics of the education system where cheating and misconduct occur frequently. Since cheating and misconduct occurs as early as the middle school years of a student, the most appropriate context to discuss ethical questions is in the primary and secondary schools, followed by the universities (Chapter 4). Finally, discussion of ethical implications should be at the annual meetings of scientific and engineering societies (Smaglik and MacIlwain, 2001) – if by then it is not too late!

Teachers and professors need to make themselves more aware of the unethical and immoral implications of cheating and misconduct. Then they need to be more prepared to inform their students about ethical and unethical issues. Subsequently, it will be possible to come to a more general conclusion at national and international levels. The ethics of science and engineering is not only a personal problem but also a collective problem that involves all scientists and all engineers.

The continuity of civilization depends on people (i.e., scientists and engineers) interacting in a genuinely ethical manner (Madison and Fairbairn, 1999, page 3). Indeed, the occurrence of unethical practices in academia and elsewhere brings to the front-stage, not only the issue of ethics, but also the need for recognition of the nature of ethics in the age of personal image being the top dog of the group, and the recipients of copious awards (Madison and Fairbairn, 1999, page 4).

Making good ethical decisions requires that the scientist or engineer has a trained sensitivity to ethical issues and a practiced method for exploring the ethical aspects of a decision and weighing the considerations that should impact the choice of a course of action. Having a method for ethical decision making in the scientific and engineering fields is, and when used regularly, such a method that becomes second nature assuming the scientist or engineer can work through ethical issues automatically without consulting the specific steps (Markkula, 2010).

References

Alcorn, P.A. 2001. *Practical Ethics for a Technological World*. Prentice Hall, Upper Saddle River, New Jersey.

Annas, J. 2006. Virtue *Ethics. In The Oxford Handbook of Ethical Theory*. D. Copp (Editor). Oxford University Press, Oxford, United Kingdom. Page 515–536.

Bergman, R. 2004. "Identity as Motivation: Toward A Theory of Moral Self." In *Moral Development, Self and Identity*. D.K. Lapsley and D. Navaez (Editors). Lawrence Erlbaum Associates, Rahway, New Jersey. Page 21–46.

Bernardi, R.A., Baca, A.V., Landers, K.S. and Witek, M.B. 2008. "Methods of Cheating and Deterrents to Classroom Cheating: An International Study." *Ethics and Behavior*, 18(4): 373–391.

Brabeck, M.M. and Brabeck, K.M. 2009. "Feminist Perspectives on Research Ethics." In *The Handbook of Social Research Ethics*. D.M. Mertens and P.E. Ginsberg (Editors). Sage Publications, Thousand Oaks, California. Pages 39–53.

Chisholm, R.M. 2008. "Libertarianism: The Case for Free Will and its Incompatibility with Determinism." In *Reason and Responsibility*. J. Feinberg and R. Shafer-Landau (Editors). Thomson-Wadsworth, Belmont, California. Page 418–425.

Davis, S.F., Grover, C.A., Becker, A.H., and McGregor, L.N. 1992. "Academic Dishonesty: Prevalence, Determinants, Techniques, and Punishments." *Teaching of Psychology*, 19: 16–20.

Furrow, D. 2005. *Ethics: Key Concepts in Philosophy*. Continuum, Press, New York.

Haydon, G. 2006. *Education, Philosophy and the Ethical Environment*. Routledge, New York.

Howard, R.A. and Korver, C.D. 2008. *Ethics for the Real World*. Harvard Business Press, Boston, Massachusetts.

Iaccarino, M. 2001. "Viewpoints: Science and Ethics." European Molecular Biology Organization, EMBO Reports, 2(9): 747–750.

Kearney, R. 1999. "The Crisis of the Image: Levinas' Ethical Response." In *The Ethics of Postmodernity: Current Trends in Continental Thought*.

G.B. Madison and M. Fairbean (Editors). Northwestern University Press, Evanston, Illinois. Page 12–23.

Kezar, A. Glenn, W.J., Lester, J., and Nakamoto, J. 2008. "Examining Organizational Contextual Features that Affect Implementation of Equity Initiatives." *The Journal of Higher Education*, 79(2): 125–159.

Kibler, W.L. 1998. "The academic dishonesty of college students: The Prevalence of the Problem and Effective Educational Prevention Programs." In *Academic Integrity Matters*. D.D. Burnett, L. Rudolf, and K.O. Clifford (Editors). Washington D. C.: National Association of Student Personnel Administrators. Pages 23–37.

Kiriakidis, S. 2008. "Application of the Theory of Planned Behavior to Recidivism: The Role of Personal Norm in Predicting Behavioral Intentions." *Journal of Applied Social Psychology*, 38(9): 2210–2221.

Kitchener, K.S. and Kitchener, R.F. 2009. "Social Science Research Ethics Historical and Philosophical Issues." In *The Handbook of Social Research Ethics*. D.M. Mertens and P.E. Ginsberg (Editors). Sage Publications, Thousand Oaks, California. Page 5–22.

Kraft, P., Rise, J., Sutton, S., and Roysamb, E. 2005. "Perceived Difficulty in the Theory of Planned Behavior: Perceived Behavioral Control or Affective Attitude." *British Journal of Social Psychology*, 44(3): 479–496.

Lillie, W. 2001. *An Introduction to Ethics*. Allied Publishers Limited, New Delhi, India.

Lincoln, Y.S. 2009. "Ethical Practices in Qualitative Research." In *The Handbook of Social Research Ethics* edited by D.M. Mertens and P.E. Ginsberg. Sage Publications, Thousand Oaks, California. Pages 150–169.

Madison, G.B. and Fairbairn, M. 1999. "Introduction." In *The Ethics of Postmodernity: Current Trends in Continental Thought*. G.B. Madison and M. Fairbairn (Editors). Northwestern University Press, Evanston, Illinois. Pages 1–11.

Marcoux, H.E. 2002. "Kansas State University Faculty Perspective, Opinions, and Practices Concerning Undergraduate Student Academic Dishonesty and Moral Development." A Dissertation Submitted in Partial Fulfillment of the Requirements for the Degree Doctor of Philosophy. Department Of Counseling And Educational Psychology College Of Education, Kansas State University, Manhattan, Kansas.

Markkula Center for Applied Ethics. 2010. *A Framework for Thinking Ethically*. Santa Clara University, Santa Clara, California. http://www.scu.edu/ethics/practicing/decision/framework.html

Moshman, D. 2004. "False Moral Identity: Self-Serving Denial in the Maintenance of Moral Self-Conceptions." In *Moral Development, Self and Identity*. D.K. Lapsley and D. Narvaez (Editors). Lawrence Erlbaum Associates, Rahway, New Jersey. Pages 83–109.

Nucci, L. 2004. "Reflections on the Moral Self Construct." In *Moral Development, Self and Identity*. D.K. Lapsley and D. Narvaez (Editors). Lawrence Erlbaum Associates, Rahway, New Jersey. Page 11–132.

NSB. 2010. Science and Engineering Indicators 2010. Report No. NSB 10-01. National Science Foundation, Arlington, Virginia.

Shrader-Frachette, K.S. 1994. *Ethics of Scientific Research*. Rowman and Littlefield Publishers, New York.

Smaglik, P. and MacIlwain, C. 2001 "Scientists Seek Solidarity in Oaths." *Nature*, 409: 971.

Smith, L.M. 1990. "Ethics, Field Studies and the Paradigm Crisis." In *The Paradigm and Dialog*. E.G. Guba (Editor). Sage Publications, Thousand Oaks, California. Page 139–157.

Stevenson, C.L. 2006. "The Nature of Ethical Disagreement." In *Philosophical Horizons*. S.M. Cahn and M. Ekert (Editors). Thomson-Wadsworth, Belmont, California. Page 284–288.

Thomas, V.G. 2009. "Critical Race Theory: Ethics and Dimensions of Diversity in Research." In *The Handbook of Social Research Ethics*. D.M. Mertens and P.E. Ginsberg (Editors). Sage Publications, Thousand Oaks, California. Page 54–68.

Walker, L.J. 2004. "Gus in the Gap: Bridging the Judgment-Action Gap in Moral Functioning." In *Moral Development, Self and Identity*. D.K. Lapsley and D. Navaez (Editors). Lawrence Erlbaum Associates, Rahway, New Jersey. Page 1–20.

Weeks R A and Kinser D L (Editors). 1994. *Editing the Refereed Scientific Journal: Practical, Political, and Ethical Issues*. IEEE Press, Institute of Electrical and Electronics Engineers, New York.

Welfel, E.R. and Kitchener, K.S. 1999. "Introduction to the Special Section: Ethics Education – An Agenda for the 90s." In *Ethical Conflicts in Psychology 2nd Edition*. D.N. Bersoff (Editor). American Psychological Association, Washington DC. Page 133–37.

Williams, B. 2006. *Ethics and the Limits of Philosophy*. Routledge, New York.

2

Scientists and Engineers

2.1 Introduction

The scientific and engineering fields are composed of educated and relatively young professionals who have the ability to apply themselves to the problems at hand, either through theory studies or experimentation. To the scientist and engineer, the outcome of this work that offers some form of gratification is: (1) completion of a project and (2) publication of the data in a journal or similar medium for distribution to one's peers. The latter gives the scientist and engineer recognition for their work.

A scientist is a person who has scientific training or who works in the field of science. On the other hand, an engineer is a person who is schooled and trained as an engineer. The practical difference between the two lies in the educational degree and the description of the task being performed by the scientist or engineer. It is widely believed that scientists explore the natural world discovering new knowledge while engineers apply that knowledge to solve practical problems, often with an eye toward optimizing cost, efficiency, or other parameters.

Research in the scientific and engineering disciplines offers the exhilaration of discovery. In addition, researchers have the opportunity to associate with colleagues who have made important contributions to human knowledge, peers who think deeply and care passionately about subjects of common interest, and students who can be counted on to challenge assumptions. Also, scientists and engineers have many opportunities to work with different people to explore new fields and broaden their expertise, especially where disciplines overlap.

However, research in any technical discipline can entail frustration and disappointment as well as satisfaction even when an experiment fails or a hypothesis turns out to be incorrect. Instead of bemoaning the outcome of the experiment, the investigator should determine if the experimental design was correct (or poor), and then determine whether the collapse of a favored hypothesis or a modified hypothesis is more sensible and logical than the previous hypothesis.

Careers in science and engineering are endeavors that can improve people's lives resulting in knowledge that all people can share. As the techniques and products of science and technology have become more central to modern society, a background in science and engineering has become essential to a majority of careers. In fact, degrees in science and engineering are becoming as fundamental to modern life as the traditional liberal-arts degree. The contributions of scientists and engineers extend beyond research and development and reach into sectors of teaching, business, industry, and government.

Graduates with bachelor's, master's, and doctoral degrees in science or engineering are increasingly forming companies, managing businesses, practicing law, formulating policy, consulting, and running for political office. They are forming global communities of common interests that transcend the differences among individuals, corporate endeavors, or nations. Anyone contemplating a career in science or engineering can maximize the potential for success in any one of several ways.

It is important to remember that each science-oriented student is unique. His success will depend on going where his particular interests lead you. If the student is exhilarated by the challenge of a new problem, puzzle, or a need or if the complexity of the natural world prompts a desire for him to understand it better, science and engineering study, rigorous though it is, will provide the tools and concepts that are needed to achieve the goals.

The individual goals of each person will determine which academic degree is most appropriate. Many baccalaureate graduates find satisfying careers in a variety of positions after the bachelor's degree. Other baccalaureate graduates, notably engineers, find that a master's degree equips them well for professional careers. For those who hope for careers conducting research and/or teaching at the higher level, a PhD will likely be required. However, no degree guarantees lifetime employment; like professionals in other fields, it is likely the graduate will change jobs and even careers during his professional lifetime, perhaps even more than once. It is important to remember that science-oriented students are not all alike, any more than all artists or all politicians are alike. Success will depend on going where your particular interests lead you, the scientist or engineer.

A young scientist/engineer must persevere until he discovers whether the rewards and compensations of a scientific or engineering life are worth the disappointments and the toil of daily professional life. Once a scientist or engineer experiences the exhilaration of making new discoveries and the satisfaction of carrying through a really difficult experiment then he may be convinced that there is no other kind of life (Medawar, 1979).

Persons educated as scientists and engineers are meant to provide service to society via their development of original ideas, which are brought to fruition in the teaching, industry, business, and government sectors. Graduate students often exceed the thinking of their professors by creating new schools of thought in their respective fields. Both the students and the professor learn from each other, often the professor gaining the most knowledge.

A cohesive education system is therefore both as important as a source of future leaders in science and engineering, and as a source of new ideas. Countries should maintain the strength of the education system to sustain the creativity and intellectual vigor that will be needed to address a growing variety of social and economic concerns.

The efficacy of most national education systems originated in a series of governmental movements that were prompted by the major role that science and technology played in the outcome of World War II. As a result, in many of these countries, government agencies assumed an important role in funding basic and applied research. Furthermore, through this form of funding, researchers at universities throughout the country become major contributors to the nation's scientific research experts. In addition, programs are conceived that would involve industry as a partner. Research in universities is essential for modern progress, but if it is lacking the focus brought on by the needs of industry, the usefulness of such research should be questioned.

A person does not need to be exceptionally talented in the scientific or engineering disciplines to be a good scientist or a good engineer; however, common sense is a critical component for achieving success. The student's potential also depends on several attributes of good character and work ethic, such as:

1. application to the task at hand,
2. diligence,

3. a sense of purpose,
4. the power to concentrate, and
5. the ability to persevere and not be cast down by adversity (Medawar, 1979).

Individual goals will determine which academic degree is most appropriate for the young person about to enter academia. Many people find satisfying careers in a variety of positions after the bachelor's degree. Others, notably engineers, find that a master's degree equips them well for professional careers. For those who hope for careers conducting research and/or teaching at the university level, a PhD will most likely be required. No degree guarantees lifetime employment. Like professionals in other fields, the young scientist/engineer might still have to change jobs and even careers during his life, perhaps more than once.

Science and engineering are not only the collection of facts but also the treatment of facts to discover their arrangement and government; from this pursuit knowledge arises and as a result its applications. The pursuit of science and engineering requires, above all things, freedom of thought, freedom of interest, and unrestricted communication. The necessity of these conditions is undeniable.

The path to a scientific or engineering career is mentally, physically, and emotionally demanding. Not everyone has the perseverance to complete years of concentrated study, but the experience of doing scientific or technical work is supremely exhilarating for those with sufficient interest and determination. There may be many teachers, mentors, or colleagues who are willing to help the merging scientist or engineer; and to assist them with overcoming difficult hurdles, therefore, enabling them to gain confidence, to expand thinking, and to work independently.

However, adventurous observation and experiment may lead down winding paths, creating new boundaries that have many possibilities of interpretation. The paths in science and technology set by the professionals before us are not merely indications of known ground, but they create a cross-correlation of many observers' experiences.

So it has been the pride and delight of scientists to rejoice in their intellectual independence; which has resulted in an astonishing knowledge of great and little achievements. Very few scientists have capitalized financially on their knowledge because to do so could restrict progress by limiting the use of available information.

The freedom and independence of science and engineering has ordained the formations of societies to enable discussion and exchange of mutual interest. The production of society journals makes carefully digested knowledge available to whosoever will read them and so knowledge becomes disseminated openly for universal use.

It is the purpose of this book to help lay the foundation for a journey into science and engineering, no matter how many turns the path might take. Just how rigorous is the path to a scientific or engineering career? Graduate study, in particular, is demanding mentally, physically, and emotionally. Not everyone has the perseverance to complete years of such concentrated study. But the experience of doing scientific or technical (sci-tech) work is supremely exhilarating for those with sufficient interest and determination. And many people will be willing to help the student along the way and assist him over difficult hurdles as to gain the confidence to think and work independently.

2.2 Definitions

In this book, the term *scientists and engineers* refers to persons who wish to pursue, or those who have already attained, a master's degree or doctor of philosophy degree (or equivalent degree) in science or engineering.

Science is regarded as inclusive of the life sciences, physical sciences, mathematics, and social science (the scientific study of human society and social relationships). *Engineering* is viewed as a field that includes all specialties such as civil, mechanical, electrical, petroleum, and computer engineering. In all cases, the system of education of scientists and engineers should be organized around realistic experience.

Education is basic to achieving national goals in two ways. First, schools and universities are responsible for producing the teachers and researchers. Investigators in academia and industry lay the groundwork for the innovations of tomorrow.

By educating students in the context of research, the systems for the education of scientists and engineers have set national standards for preparing scientists and engineers for research careers in academia, government, and industry. Furthermore, by attracting outstanding students and faculty members (hopefully who have some understanding of the non-academic world) national systems have, to some extent, benefited from an infusion of both talent and ideas.

The increase in scientific and technological knowledge and the ways in which that knowledge is applied are fundamental to the pursuit of many general national objectives, including developing new technologies and industries, combating disease and hunger, reducing environmental pollution, developing new sources of energy, and maintaining the competitiveness of national industry.

Supposedly, at least in Western societies, science and engineering are regarded by many as the ultimate arbiters of objective truth (Ryan, 2002; Brown, 2003). However, there must be an intrinsic accountability that arises in the experimental. Experiments need to be repeated, discrepancies noted, and questions asked but while errors can be attributed to mistakes there is often no contemplation of fraud, either as an afterthought or deliberate. To get beyond such a hurdle, it is always advisable to encourage scientists and engineers to double check, triple check, and check their work *ad nauseam*.

Furthermore, any scientist or engineer who is requested to be a coauthor should ignore the data in next-to-final draft before publication (after the data have been massaged to look presentable) and check the original data.

Persons educated as scientists and engineers are meant to provide service to society via their development of original ideas, which are brought to fruition in teaching, industry, business, and government. Graduate students often go beyond the thinking of their professors and create a new generation of science and engineering thought. The student learns from the professor, but the professor, if he will admit it, also learns from the student.

A cohesive system of science and engineering education, therefore, is important both as a source of future leaders in science and engineering, and as a source of new ideas. Countries should maintain the strength of this system to sustain the creativity and intellectual vigor that will be needed to address a growing variety of social and economic concerns.

The efficacy of most national education systems originated in a series of governmental movements that were prompted by the major role that science and technology played in the outcome of World War II. As a result, in many of these countries, government agencies, assumed an important role in funding basic and applied research. Furthermore, through this form of funding, researchers at universities throughout the country became major contributors to the nation's scientific research expertise. In addition, programs were conceived that would involve industry as a partner. Research in universities is fundamental, but if such research lacks the focus brought on by the needs of industry, the usefulness of it must be questioned.

The dual role of the university-industry partnership was designed to benefit national goals by educating students through the active conduct of cutting-edge research and its application to industry.

With an increased spread of information and the global workforce, guidelines for the theme of the book, "Educating

Scientists and Engineers" can be obtained instantaneously around the world. This brings into question the impact of globalization. Jane Knight described globalization as, 'the flow of technology, economy, knowledge, people, values, and ideas across borders." Globalization affects each country in a different way due a nation's individual history, traditions and priorities.

For the education of scientists and engineers, globalization is an inescapable reality. It increases the flow of students from all parts of the world to get a better understanding of international conditions. Globalization is also visible in the current migration of engineering and science graduates from one country to another in search of better jobs and more affordable learning opportunities. This has created a condition of *brain drain* in some of the lesser developed countries such as India and the Republic of Trinidad and Tobago.

In general, there are two ways to contend with the forces of globalization. First, universities must adopt a defensive approach in order to protect its turf from outside educational forces that could have negative effects on teaching and learning. Second, universities must embrace globalization in an age of new opportunities to advance the outcome of quality scientists and engineers; thereby offering increased value to expertise and talent for national development.

2.3 Scientific Disciplines

Science (Latin *scientia*: knowledge) refers, in the broadest sense, to any systematic knowledge or practice. In a more restricted sense, science refers to a system of acquiring knowledge based on the scientific method, as well as to the organized body of knowledge gained through such research.

Fields of science are commonly classified along two major lines: natural sciences, which study natural phenomena (including biological life), and social sciences, which study human behavior and societies. These groupings are empirical sciences, meaning that the knowledge must be based on observable phenomena and capable of being proved through experimentation by other researchers working under the same conditions.

Mathematics, which is sometimes classified within a third group of science (called formal science), has both similarities and differences with the natural and social sciences. It is similar to empirical sciences in that it involves an objective, which is a careful and systematic study of an area of knowledge. It is different because of its method of verifying its knowledge and does not use empirical methods. Formal science, which also includes statistics and logic, is vital to the empirical sciences; although one should not always place implicit faith in the use of statistics as there may be pitfalls in the use of such methods (Huff, 1954; Gibilisco, 2004). Major advances in formal science have often led to major advances in the physical and biological sciences. The formal sciences are essential in the formation of hypotheses, theories, and laws, both in discovering and describing how things work (natural sciences) and how people think and act social sciences).

The history of science is marked by a chain of advances in technology and knowledge that complement each other. Technological innovations bring about new discoveries, bred by other discoveries, which inspire new possibilities and approaches to longstanding science issues. Investing capital in science and engineering is critical to ensuring a high quality of life.

Scientists and engineers are at the forefront of the development of scientific and technological innovations. The primary objectives of these professionals are to create and develop novel research that can be used to solve problems for both the population at large as well as individual entities such as companies.

Well into the 18th Century, science and philosophy were roughly synonymous. In fact, the preferred term for the study of nature was often *natural philosophy*, while English speakers most typically referred to the study of the human mind as *moral philosophy*. By the early 1800s, science had begun to separate from philosophy, though it often retained a very broad meaning. In many cases, it stood for reliable knowledge about any topic. It was often linked to a set of well-defined laws, not just of nature but among any phenomena. Over the course of the 19th Century, however, there was an increased tendency to associate science with the natural world (that is, the non-human world). This move sometimes left the study of human thought and society (social science) in a scientific limbo by the end of the nineteenth century and into the twentieth century.

Over the 1800s, many English speakers were increasingly differentiating science from all other forms of knowledge in a variety of other ways as well. For instance, the scientific method (i.e., used to explain the events of nature in a reproducible way, and to use these reproductions to make useful predictions) was almost unused during the early part of the nineteenth century. It became widespread only after the 1870s, though there was scarcely total agreement about just what it entailed. Similarly, discussion of scientists and engineers as a special group of people did not always emphasize the attributes of such education. Whatever was actually meant by them, such terms ultimately depicted science as something deeply distinguished from all other realms of human endeavor.

By the 20th Century, the modern notion of science as a special brand of information about the world, practiced by a distinct group and pursued through a unique method was essentially in place. Over the 1900s, links between science and technology also grew increasingly strong. By the end of the century, it is arguable that technology had even begun to eclipse science as a term of public attention and praise.

A *scientific discipline* is a particular branch of science. However, there is more to the definition.

In the early nineteenth century, anyone involved in science was generally labeled a *natural philosopher.* As the century progressed, science evolved into different disciplines and then into its current sub-disciplines that are now known collectively as *science.* At that time, science moved into the universities to avoid being dominated by technologists, who at the same time were enjoying a high degree of social status due to their successes during the industrial revolution. As science moved into the universities, German academics categorized the study of nature into physics, chemistry, biology, and geology (Grau, 1988). The academics also organized their university science administrative units the same way. Other universities worldwide followed Germany's lead and used the same categories to organize their own newly established science departments.

Towards the end of the nineteenth century, the high school science curriculum was developed in Europe and North America. It was only natural that this curriculum would be organized around the university's administrative units of physics, chemistry, biology, and geology. The *sub-disciplines* of science were established in the high school curriculum and they are now over one hundred years old.

Although science evolved into simple batch discipline of chemistry (also included biochemistry), physics, biology (composed of botany and zoology), and scientific disciplines are now much more diverse.

The current definition for a scientific sub-discipline is an academic discipline, or field of study, it is a branch of knowledge which is taught or researched at the university level. Disciplines are further defined and recognized by the journals in which research is published, and the learned societies and academic departments or faculties to which their practitioners belong. The *fields of study* usually have several sub-disciplines or branches, and the distinguishing lines between these are often both arbitrary and ambiguous.

In medieval Europe, there were typically four faculties (or colleges) in a university: *theology, medicine,* law, and arts. Arts had a somewhat lower status than the other three faculties. Modern university faculties have their beginnings in the mid- to late-nineteenth century popularization of universities, when the traditional curriculum was supplemented by non-classical languages, non-classical and literature, and by disciplines such as chemistry engineering, physics, and biology.

In the early twentieth century, new disciplines such as education, sociology, and psychology were added. In the 1970s and 1980s, there was an explosion of new disciplines focusing on specific themes, such as *women's studies* and *environmental studies* as well as a host of studies related to the ethnic origins of the various peoples of the world. Finally, the visibility of such interdisciplinary scientific fields as biochemistry and geophysics increased, as their contribution to knowledge became widely recognized.

Currently in academia, it is a growing practice to incorporate fields of study that are created by extending the ideas, theories, and methods of more traditional disciplines. Also, new times and revolutionary thinking can enhance or renew existing disciplines, or even create new disciplines altogether.

2.4 Engineering Disciplines

Engineering is the discipline of acquiring and applying scientific and technical knowledge to the design, analysis, and/or construction of works for practical purposes. Alternatively, engineering is also "the creative application of scientific principles to design, develop structures, machines, apparatus, manufacturing processes, or works utilizing them singly or in combination; or to construct or operate the same with full cognizance of their design; or to forecast their behavior under specific operating conditions; all as respects an intended function, economics of operation and safety to life and property." (ECPD, 1941)

One who practices engineering is called an engineer, and those licensed to do so have formal designations such as Professional Engineer (PE) or Chartered Engineer (CE). The broad discipline of engineering encompasses a range of specialized sub-disciplines that focus on the issues associated with developing a specific kind of product, or using a specific type of technology.

Just as scientific disciplines can be said to span the alphabet, engineering disciplines also are many-fold and offer a wide choice of options. New sub-disciplines such as bioengineering, which combines biology and engineering, are thriving. Bioengineers work closely with biologists and medical doctors to develop medical instruments, artificial organs, and prosthetic devices.

The history of the concept of *engineering* stems from the earliest times when humans began to make clever inventions, such as the wheel, the pulley, and the lever. The exact meaning and origin of the word *engineer*, however, is a person occupationally connected with the study, design, and implementation of engines (Latin: *ingenium*, innate quality), especially mental power for the purpose of a clever invention. Hence, an engineer, essentially, is someone who makes useful or practical inventions.

From another perspective, a now obsolete meaning of engineer, dating from 1325, is "a constructor of military engines." Engineering was originally divided into military engineering, which included construction of fortifications as well as military engines, and civil engineering, involved in non-military projects, such as bridge construction.

One of the largest branches of engineering, civil engineering is a field that deals with buildings, bridges, dams, roads, and other structures. Civil engineers (who evolved from the ancient military engineers) plan, design, and supervise the construction of facilities such as high-rise buildings, airports, water treatment centers, and sanitation plants. Civil engineers will need to design the special rail beds for the magnetic levitation trains of tomorrow.

With the rise of engineering as a profession in the nineteenth century the term *engineering* became more generally applied to fields in which mathematics and science were applied to the industrial applications. Similarly, in addition to military and civil engineering the fields then known as the *mechanical arts* became incorporated into engineering.

2.5 Expert Witness

An *expert witness*, in the context of this book, is scientist or engineer who is called upon and agrees to present testimony to a court or other legal body.

Scientists and engineers increasingly are asked to serve as consultants who provide expert testimony in adversarial or potentially adversarial contexts (Speight, 2009). The focus of the case might be on the past, as in explaining the causes of accidents, malfunctions, and other events involving technology. Usually the scientist or engineer is hired by one adversary in the dispute, and that raises special ethical concerns about their proper roles.

Because of the complexity of science and engineering, the court system must rely on experts who earnestly try to be impartial in identifying and interpreting complicated data. Ideally, expert witnesses would be paid by the courts, rather than opposing attorneys, in order to counter potential biases. In practice, the high costs require that parties to the disputes pay for expert witnesses.

Some people are uncertain of the function of expert witnesses; do they play the role of an impartial communicator of truth or as the *hired gun*, who is a witness paid to tell only one side of the story? The role of the expert witness is to present the truth to the courts and not to act as an advocate for one side or the other and certainly not to act as a partial witness telling only that part of the truth that supports the arguments of the side paying him. Some scientist and engineers believe that the role of the impartial analyst (who states and assesses facts) and the advocate (who makes recommendations about responsibility and preferable options) is not altogether precise. This is a false belief.

Although the scientist or engineer might feel responsibilities to the attorneys who hire them, they have obligations to represent their qualifications accurately, to perform thorough investigations, and to present a professional demeanor when called to testify in court. Equally important, they have a responsibility of confidentiality, just as they do in other consulting and other related jobs. The guiding rule is to present the truth to the court. The truth is based on facts and the facts presented to the court are based on the ethics of the expert witness. Any scientist or engineer who cannot abide by such a rule is well advised not to take on the role of an expert witness or, if taking on the role, better understand the meaning of the word *perjury*, and be prepared to suffer the consequences.

It would be unethical for an expert witness to divulge the contents of their investigations to the opposing side of a controversy until required to do so by the courts or by the attorney who hired them. Most important (as previously mentioned) when called as a witnesses the scientist or engineer is not required to volunteer evidence favorable to the other side. They must answer questions truthfully, but it remains the responsibility of the attorney for the opposing side to ask pertinent questions.

Codes of ethics have only recently begun to clarify the roles of scientist and engineers in legal adversarial contexts, and as a result there has been some light shed upon the appropriate role of the scientist or engineer (Speight, 2009). It follows from the various codes of ethics that scientists and engineers must not take on the role of the *hired gun* who engages in distorting facts according to who pays the consulting fee.

Merely being paid by one side can exert a slight bias, which might influence the investigations, testimony, and even the presentation of qualifications of the scientist or engineer. The bias would increase substantially if the scientist or engineer was hired on the basis of contingency fees paid only if the case is won. The various codes of ethics notwithstanding, it is perceived that contingency fees in adversarial contexts would tend to bias the judgment of the expert witness.

In summary, as expert witnesses, the scientists and engineers should be completely impartial. Not only should the scientist or engineer conscientiously avoid any taint of bias and favoritism, but they should avoid any form of advocacy. The role of the expert witness is to identify all options and analyze the factual implications of each option (Speight, 2009).

2.6 Professionalism

Science and *engineering* are the disciplines of acquiring and applying scientific and engineering to the design, analysis, and/or construction of works for practical purposes.

A profession is any occupation that provides a means by which to earn a living. In the sense intended here, the scientific and engineering professions are those forms of work involving advanced expertise, self-regulation, and honesty (Martin and Schinzinger, 2005). Scientific and engineering professionals play a major role in setting standards for admission to the profession, drafting codes of ethics, enforcing standards of conduct, and representing the profession to others. Professionals should maintain high ethical standards and to do so brings the recognition traditionally associated with the word *profession*.

With the rise of science and engineering as professions, the terms became more generally applied to fields in which the sciences and mathematics were applied to the industrial applications. Similarly, in addition to military and civil engineering the fields then known as the *mechanic arts* became incorporated into engineering.

Professionalism entails a multiplicity of tasks and a variety of new roles; not all individuals occupying these roles of trust have been adequately prepared for and socialized to them. Actions are often collective, i.e., via team approaches to problem posing and problem solving, which can undermine individual responsibility. Indeed, the importance of recognizing the role of the society in contributing to incidences of research misconduct was noted during conference discussions (Chalk et al., 1980; Chalk, 2005). All of these potentially conflicting factors may make it difficult for a researcher to know with confidence what is ethically expected of him or her (Gorlin, 1986; Davis, 2002).

From another perspective, a now obsolete meaning of engineer, dating from 1325, is "a constructor of military engines." Engineering was originally divided into military engineering, which included construction of fortifications as well as military engines, and civil engineering, involved in non-military projects, such as bridge construction.

One of the largest branches of engineering, civil engineering is a field that deals with buildings, bridges, dams, roads, and other structures. Civil engineers (who evolved from the ancient military engineers) plan, design, and supervise the construction of facilities such as high-rise buildings, airports, water treatment centers, and sanitation plants. Civil engineers will need to design the special rail beds for the magnetic levitation trains of tomorrow.

With the rise of science and engineering as professions, the terms became more generally applied to fields in which the sciences and mathematics were applied to the industrial applications. Similarly, in addition to military and civil engineering the fields then known as the *mechanic arts* became incorporated into engineering.

Chemical engineering, like its counterpart *mechanical engineering,* developed in the nineteenth century during the Industrial Revolution. Industrial scale manufacturing demanded new materials and new processes. By 1880 the need for large scale production of chemicals was such that a new industry was created, dedicated to the development and large scale manufacturing of chemicals in new industrial plants. The role of the chemical engineer was the design of these chemical plants and processes such as processing and treating of liquids and gases. Many chemical engineers work with petroleum and plastics, although both of these are the subject of independent disciplines. The sub-discipline *environmental engineering* also applies to certain areas of chemical engineering, such as pollution control.

Just as scientific disciplines can be said to span the alphabet, engineering disciplines also are many-fold and offer a wide choice of options. New sub-disciplines such as bioengineering, which combines biology and engineering, are thriving. Bioengineers work closely with biologists and medical doctors to develop medical instruments, artificial organs, and prosthetic devices.

The scientist and/or engineer may be required to establish that he has performed better insofar as the work can be established as better than that of any predecessor. Then the definition of "better" becomes an issue that detracts from the real issue – getting the best out of scientist and engineers. Put simply, "better" can mean: "a process for producing an improved product," "completion of a project under budget," or anything in between these two limits.

Therefore, the fundamental basis for deciding on achievement may be lost because of a complete misunderstanding of the nature of the work and its importance to the university or to the company. The fell effect of the contribution is lost in the argument of word definitions.

For the purposes of evaluating a scientist or engineer, there should be two critical objectives of an evaluation and these are recognition of the significance of the contribution and any ensuing objectives and the magnitude and significance of the impact of the contribution.

Suddenly, the young scientist or engineer sees that his work is being recognized as meaningful. Morality develops and the young professional is satisfied, for the moment! However, the issue remains of the means by which the impact of the contribution can be assessed.

The scientific and engineering fields are composed of educated and relatively young professionals who have the ability to apply themselves to the problems at hand, either theory studies or experimentation. To the scientist and engineer, the outcome of this work that offers some form of gratification is the completion of a project and the publication of the data in a journal or similar medium for distribution to one's peers. The latter gives the scientist and engineer recognition for their work.

In general, the professionals who are biased towards theory tend to produce data that are often abstract and the intellectual contribution is expressed in the form of theories with proof. As a result, publication on the proceedings of a conference may be the only outlet for their efforts after which publication in a reputable journal may be possible but only with considerable efforts, for various reasons, may not be possible at all. For the non-academic scientist and engineer, there is the medium of publication of their field's material compiled as a *company report*. This can be a worthwhile method for circulating one's work throughout the company. But, the importance of the work to the young scientist and engineer can, again, be diminished and the names of a supervisor and any other persons higher up the food chain should be included as co-authors.

Publication of data in the proceedings from a conference often results in a shorter time to print. This follows from the opportunity to describe completed or partly completed work before peer scientists and/or engineers and to receive a more complete review than the type of review that is typical for a journal. At a conference, the audience asks general and specific questions to the presenter that often provides recommendations for further work or a new line of investigation. Overall, this will help the presenter to finalize the document for publication in the proceedings (where the proceedings are published post-conference). On the other hand, one has to wonder of journal reviewers really pay attention to the salient points of the potential publication or do they merely look for errors in style and grammar. An answer that several readers may relate to is "all of the above." However in many academic reviews, statements are made that publication in the proceedings of a prestige conference is inferior to the publication in a prestige journal; without realizing or admitting that data presentation and publication, in many conferences are superior to an established journal.

Yet, publication of research data is not an open form of recognition for all scientists and engineers. Scientists and engineers employed in industry may be prevented from publishing their work because of a company policy related to proprietary material, which is a justified reason, or an arbitrary decision by a supervisor or a member of the company review committee, which is not a justified reason.

On the other hand, in academia, the young professional enters a department at the Assistant Professor grade. At this level, the Assistant Professor has little choice in terms of choosing teaching assignments and has administrative work thrust upon his shoulders, while the older tenured members of staff have the right to refuse such work without fear of reprisal. Yet, this is not the reason behind tenure.

Tenure was introduced to protect academic freedom in educational settings from the whims of politics – whether this is in the form of meddling from the outside or from the inside. Tenure was thus introduced to preserve academic autonomy and integrity because it was recognized that this was beneficial for the state, for society, and for academia. Tenure was not designed to allow faculty to refuse work!

A professor who holds tenure is virtually an impregnable fortress and cannot (without considerable effort and expense) be dismissed from his appointment. The appointment is essentially for life. Tenure has come under attack over the past three decades by those who want a more business-like approach to universities, including ending tenure, accountability, performance review, audits, and performance-based salaries (Hacker and Dreifus, 2010; Taylor, 2010).

In addition, the young Assistant Professor also has to acquire research funding and may even have to pass his reports/papers through a review committee prior to publication. This review committee will be made up of senior members of staff who, for many reasons that are often difficult to follow, can give the young professor a glowing performance report or a report that is somewhat less than glowing. It is at this time, if the latter is the case, that the young professor can feel that he is suffering rejection by one's colleagues.

The educated young professional scientist and engineer wonders if he is merely a pair of hands (for a overbearing supervisor, an overbearing department head or jealous colleagues) who is not supposed to be given credit for the ability to think and solve a problem. Performance suffers and, with repeated negativism towards publication, the young professional starts to lose interest in the organization.

Lack of recognition for hard and intelligent work is a killer and getting the best out of any such scientists and engineer becomes an impossible dream.

Publication of data in the proceedings from a conference often results in a shorter time to print, resulting in a partly completed review that is typical for a journal. At a conference, the audience asks general and specific questions to the presenter that often provides recommendations for further work or a new line of investigation. Overall, this will help the presenter to finalize the document for publication in the proceedings (where the proceedings are published post-conference). On the other hand, one has to wonder if journal reviewers really pay attention to the salient points of the potential publication or do they merely look for errors in style and grammar. An answer that several readers may relate to is all of the above. However in many academic reviews, statements are made that publication in the proceedings of a prestigious conference is inferior to the publication in a more prestigious journal, without realizing or being willing to admit that in relation to data presentation and publication, many conferences are superior to an established journal.

For the purposes of evaluating a scientist or engineer, there should be two critical objectives of an evaluation; these are (1) recognition of the significance of the contribution, and (2) the magnitude and significance of the impact of the contribution.

Finally, in addition to the traditional business-like approach, many universities would do well to adopt one of the most widely printed and quoted statements of business ethics in the world, *The 4-Way Test* from Rotary International (Dochterman, 2003). The *4-Way Test* was created by Rotarian Herbert I. Taylor in 1932, when he was asked to take charge of the Chicago-based Club Aluminum Company, which was facing bankruptcy. Taylor looked for a way to save the struggling company mired in depression-caused financial difficulties. He drew up a 24-word code of ethics for all employees to follow in their business and professional lives. The *4-Way Test* became the guide for sales, production, advertising, and all relations with dealers and customers, and the survival of the company was credited to this simple philosophy. The *4-Way Test* was adopted by Rotary in 1943 and has been translated into more than 100 languages and published in thousands of ways. The message is known and followed by all Rotarians and is as follows:

1. Is it the TRUTH?
2. Is it FAIR to all concerned?
3. Will it build GOODWILL and BETTER FRIENDSHIPS?
4. Will it be BENEFICIAL to all concerned?"

These four questions can act as a guide to dealing within academia and they also afford a means of consolidating one's ethical beliefs; whether it is in science, engineering, academia, or business.

References

AAAS. 2000. *The Role and Activities of Scientific Societies in Promoting Research Integrity. A Report of a Conference*, American Association for the Advancement of Science. U.S. Office of Research Integrity, Washington, DC. http://www.aaas.org/spp/dspp/sfrl/projects/integrity.htm

Alcorn, P.A. 2001. *Practical Ethics for a Technological World*. Prentice Hall, Upper Saddle River, New Jersey.

Braithwaite, O, and Baxter, L.A. (Editors). 2006. *Engaging Theories in Family Communication: Multiple Perspectives*. Sage Publications Inc., Thousand Oaks, California.

Brown, T.L. 2003. *Making Truth: Metaphor in Science*. University of Illinois Press, Champaign, Illinois.

Chalk, R., Frankel, M.S., and Chafer, S.B. 1980. *Professional Ethics Activities in the Scientific and Engineering Societies*. American Association for the Advancement of Science, Washington, DC.

Chalk, R. 2005. *AAAS Professional Ethics Project: Professional Ethics Activities in the Scientific and Engineering Societies*. American Association for the Advancement of Science, Washington DC.

Davis, M. 2002. *Profession, Code, and Ethics*. Burlington, VT: Ashgate Publishers, Burlington, Vermont.

Dochterman, C. 2003. "The ABCs of Rotary." *Rotary International*. Evanston, Illinois.

http://www.rotary.org/en/aboutus/rotaryinternational/guiding principles/Pages/ridefault.aspx

ECPD. 1941. "The Engineers' Council for Professional Development." *Science*. 94(2446): pp. 456.

Gibilisco, S. 2004. *Statistics Demystified*, McGraw-Hill, New York.

Gorlin, R.A. (Editor). 1986. *Codes of Professional Responsibility*. Bureau of National Affairs, Washington, DC.

Hacker, A., and Dreifus, C. 2010. *Higher Education? How Colleges are Wasting our Money and Failing our Kids – And What We Can do About It*. Times Books, MacMillan, New York.

Huff, D. 1954. *How to Lie with Statistics*. W.W. Norton and Company Inc., London, United Kingdom.

Levine, F.J., and Iutcovich, J.M. 2003. "Challenges in Studying the Effects of Scientific Societies on Research Integrity." *Science and Engineering Ethics*. 9: 257–268.

Martin, M.W., and Schinzinger, R. 2005. *Ethics in Engineering 4th Edition*. McGraw Hill, New York.

Medawar, P.B. 1979. *Advice to a Young Scientist*. Alfred P. Sloan Foundation, Library of Congress, Washington, DC.

OSTP. 1999. "Proposed Federal Policy on Research Misconduct to Protect the Integrity of the Research Record." *Office of Science and Technology Policy, Executive Office of the President*. Federal Register, 64(198): 55722–55725.

Ryan, J.F. 2002. "Fraud." *Today's Chemist at Work*. American Chemical Society: Washington, DC. Volume 11(11): 9.

Speight, J.G. 2009. *The Scientist or Engineer as an Expert Witness*. CRC Press, Taylor & Francis Group: Boca Raton, Florida.

Taylor, M.C. 2010. "Crisis on Campus." *Crisis on Campus*. A Bold Plan for Reforming our Colleges and Universities. Knopf, New York.

UNESCO. 2006. Proceedings. Forum on the Code of Ethics for Scientists and Engineers Korean National Commission for UNESCO. May 26.

3

The Psychology and Philosophy of Ethics

3.1 Introduction

Ethical issues (Chapter 1) permeate every stage of the research process from the provision of a title to the study and the analysis of the data (Chapter 8). In fact, in order to study such unethical behavior in any science (including social science and political science, two sciences that are not usually based on the same forms of data gathering and data workup as the hard sciences) or engineering are a range of ethical issues emerging in the fields of qualitative and quantitative research. This has been and remains so because quantitative research is rooted in rationality and objectivity, and reflection can be used to correct/evaluate and logic of analyses done. Meanwhile, qualitative approaches to data collection are more personalized and allow for expressions of values, beliefs, motivations, emotions in sharing of information.

Furthermore, and in addition to the ethical responsibilities of researchers, respondents also have ethical responsibilities.

More often than not respondents do not breach their ethical commitments, spoken or unspoken. Researchers, for several reasons, may or may not adhere to their personal and/or professional ethics.

Ethics forms a major classification of philosophy and is a study of values and customs of a person or a group (Chapter 1). It deals with the analysis and application of concepts such as right or wrong, good and evil and clear distinction of responsibilities. Professional ethics refers to ethics specifically concerned with human character, acceptable behavior and conduct. Ethical behavior is something that goes beyond simply obeying a set of rules and regulations, it is about committing yourself to do and act according to what is right, cognizant of your own conscience. To put it simply, professional ethics concerns one's behavior, conduct and practice when carrying out professional work; it could be any profession such as consulting, research or writing. Most professional bodies set a code of conduct that is to be followed by its members such as doctors, accountants, lawyers to name a few. It is assumed that the members accept the adherence to these codes or rules, including restrictions that apply. At the same time, no two codes of ethics are identical. They vary on the basis of cultural group, profession or discipline.

Ethics in terms of philosophy, is often referred to as morality – the *right* or *wrong* of any action taken or will take place. Ethics is used to formulate judgment on any standards that are proposed for scientists and engineers to follow. Furthermore, ethics is often broken down into three main categories, as was mentioned in an earlier chapter: meta-ethics, normative ethics, and applied ethics.

Meta-ethics is the study of origin of ethical concepts and the name implies that it encompasses a whole concept of ethics. Meta-ethical issues give rise to such questions relating to the origin(s) and application(s) of human rights.

Normative ethics are the principles established that guide or regulate human conduct and are often what society considers the *norm*. Normative ethics are the guide for *proper behavior* that society sets as their standard. The golden rule is of normative ethics is: "do unto others as you would have them do to you" and NOT "do unto others before they do to you!"

Applied ethics is the study of specific problems or issues with application of *normative ethics* and/or *meta-ethics*. Often, applied ethics may involve political or social questions, but always involve some moral aspect.

Application of these three principles should help the scientist or engineer to decide the correct path to take when uncertainty arises. It is completely untrue that they cannot be responsible, cannot be held to be responsible for their actions, or cannot control what they do or what they choose to do. Indeed, scientists and engineers must be held responsible for their actions (or for their omissions). There are too many occasions when academics refuse to accept *responsibility* but will forge ahead, in order to be given unrestricted authority!

It is often claimed or acknowledged that scientists and engineers might not be responsible for their actions if what they do is the result simply of some chance, totally unexpected, unwilled, random, unexplainable, or unpredictable occurrence that takes place accidentally in their mind or body. Thus, if an act or choice is the result of forces over which the scientist or engineer had no control in the beginning then he should not be held responsible. In addition, compulsive behavior, which is unaffected by choice, can be an example of behavior which is the result of organic causes over which the scientist or engineer had no control and for which he, or it is not responsible.

If a choice or an action is the inescapable consequence of forces beyond the scientist or engineer's control, ethical principles and moral reasoning would not actually show what was right or wrong (in those cases). They would have no effect at all regarding indeterminate, chance behavior and would not be reasons for behaving in certain ways but would be causes contributing to a scientist or engineer behaving that way.

However, the ability of scientist or an engineer to act freely is not to act either compulsively (determinism) or by chance (indeterminism) but to act in regard to an informed, rational or reasoned choice, which can be examined for its reasonableness and objectiveness. This does not dismiss emotions or sensations, as some would hold, since these can be taken into account by reason.

Scientists and engineers must be held responsible for any choice they make that they could have made differently (and for any resulting action) they did that they could have done differently, even though they may not have made the choice rationally or objectively. Irrational choices, which are neither accidental nor the result of uncontrollable forces, make the scientist or engineer responsible for his actions though they may not show responsibility in behavior or decision-making. Although there may be forces over which the scientist or engineer has no control, not all wrong acts or bad choices are the result of neither inescapable forces nor forces that the scientist or engineer people could not have overcome.

Even if the personal character traits of a scientist or engineer cause the outcome of the choice, that person is still responsible for his actions. How the scientist or engineer made the choice is irrelevant.

3.2 Ethical Responsibilities in Research

Research is an activity enabling us to test some hypotheses or conclusions and contribute to knowledge"

(Shrader-Frechette, 1994) or *"the process of making and proving claims"* (Altman, 1997).

Research ethics guides researchers about the conduct necessary when carrying out studies and data workup or data manipulation (Chapter 8). However the term *data manipulation* has a ring of untruth about it! Research ethics regulations have traditionally focused on informed consent, breaches of confidentiality, stress, injury, coercion, invasion of privacy and deception.

Whether or not researchers conduct scientific research, they have an implicit obligation to society as a result of training and education that they had received (Shrader-Frechette, 1994). Complete objectivity in research is impossible because human beings cannot be completely objective with respect to the exact margin of error, choice of statistical test, sample selection, research designs, data interpretations, assumptions and theories. Most of the ethical issues arise with respect to methodological value judgments and such value judgments should be specified even if they are defensible (Shrader-Frechette, 1994). Scientific results must also be presented in a manner that would avoid future misuse or misinterpretation.

Membership in a profession carries with it an implicit commitment to pursue the welfare of the profession. This is partly done by avoiding hasty, unconfirmed statements, incomplete analyses and by speaking out about these in the studies of peers, thus the significance of peer reviews. This is why many journals have stipulations to deal with fraud and may require researchers to place their raw data in a special archive (Shrader-Frechette, 1994). However different research applications often carry different degrees of risk for the public and, as such, researchers must aspire to high standards of reliability and validity in order to minimize damaging implications. This raises concerns for knowledgeable and ethical objectivity.

Studies on scientific misconduct has found that there are several categories of people who may engage in unethical practices, deliberately or not. These are:

1. new faculty/junior scientists/junior engineers who have not been properly mentored,
2. individuals seeking promotion or tenure,
3. those who like to see their name in print (preferable not the investigative press), and
4. those who lack clear thought about the consequences and potential dangers of cheating in science and engineering.

Some examples of ethical issues in research are:

1. Failing to keep important analysis of documents of a period of time.
2. Failure to maintain complete records of findings.
3. Seeking the status of co-author without making a significant contribution to the article.
4. Not allowing one's peers access to data collected and analyzed especially after the article was published.
5. Exploiting research assistants without acknowledging their assistance.
6. Bias in sampling.
7. Continuing studies until a point of statistical significance is reached, even though the statistics are faulty.

Ethics, it has been established, is concerned with what should and should not be done and this is one of the requirements of a profession. Professional ethics constitute standards that are widely accepted within the profession (Schwartz, 2009). Generally the stipulations of ethical associations worldwide emphasize

1. high technical standards,
2. a certain range of abilities, skills and cultural knowledge,
3. integrity, honesty and respect for people, and
4. responsibility for the well being of others.

These standards should be borne in mind when developing, carrying out and reporting research results (Wolf et al., 2009). Furthermore, ethical concerns of researchers often emanate from their awareness of the entities or communities or organizations that they represent, from attempting to be neutral, or from holding on to a specific set of principles. The assessment of the research in particular would focus on the approach to the study, degree of accuracy and the accuracy of the reporting of results (Wolf et al., 2009).

With respect to experimental research, it has been established that partial success in identifying cause-and-effect relationship, such as the researcher's role in decision-making and the contribution in reducing the cost of wrong decision-making, must continue to be valued. Once this approach provides the best possible answer in the circumstances, then it is doing what good ethics requires. Ethical concerns persist however with respect to risks and benefits and decision about which causal relation is more important to be investigated (Mark and Gamble, 2009).

Rigorous research, by unearthing the truth, may leave behind social chaos, breakdown and conflict. At the same time, if such research glosses over issues and unearths partial truths, especially if it is consciously done, that is unethical. Unethical development research is that which has a covert goal, peripheralizes the voices of participants, has little transparency, did not get informed consent, is not context sensitive, is insensitive to the power relationships that influence responses (Brydon, 2006). This challenge also applies to research on various aspects of family life: marriage, cohabitation, sibling ties, father involvement, inter-ethnic relations, social class and family status, religious belief, parenting and others.

The ethical issues connected with publication, authorship, a willingness to share data and the accompanying conflicts of interest are myriad (Chapter 8) (Brown and Hedges, 2009). Data sharing is critical in those aspects of research that have implications for solving problems in science and engineering. Ethical issues also surface if and when a researcher ignores measurement errors (because the data fit or do not fit his theory). In fact, measurement errors (deliberate and unconscious) can occur in quantitative research with respect to test development (e.g. item-to-total correlations, item means, test-retest reliability, factor analysis, residual analysis, validity testing, scale development) and research designs (e.g. strength of tests of hypothesis, data collection) (Viswanathan, 2005).

In considering ethical issues in science and engineering, a distinction is often made between morals and ethics. When such a distinction is made, the term *morals* is assumed to refer to generally accepted standards of *what is right and what is wrong*. The term *ethics* is assumed to refer to the principles which appear in a code of professional ethics. However, the terms *moral philosophy* or *moral theory* generally refer to a set of abstract moral principles and it is often considered appropriate or more practical to use the words interchangeably. Both of the terms refer to standards of right conduct and the judgments of particular actions as right or wrong by those standards.

Moral and ethical statements should also be distinguished from laws. The fact that an action is legally permissible does not establish that it is morally and ethically permissible. Just as legality does not imply morality, illegality does not imply immorality. It would be illegal to introduce very small amounts of a chemical into the atmosphere if doing so violates environmental standards, but there might make a philosophical argument that there are cases in which it is not immoral to do so because the environmental standards are too strict and fail to balance costs and benefits in a rational way.

The scientist or engineer is well advised to recue himself from such arguments that enter the realms of philosophy and psychology. Anyone not skilled in either of these two mental areas of scholarship will surely be on the losing side of the argument.

Ethics is a very relevant area in the study of psychology as ethical values on what is wrong and what is right relate directly to the moral standing of scientists and engineers in society. Ethical standards are closely associated with moral standards although morality is more individualistic and moral standards could vary between cultures. Ethical standards are, however, more general as they depend on our basic human nature and human values. Ethical values are more human and thus more about psychological dynamics than the moral values. Yet ethics is considered as a branch of moral philosophy.

When considering the *psychology of ethics* it is important to distinguish between ethics and morality and the *psychology of ethics* would be more about values of being human whereas *moral psychology* specifically deals with questions of morality. Moral psychology or psychology of morality is thus considered a part of the broader psychology of ethics. Ethics deals with morality as well as questions of right and wrong, moral and immoral, virtue and vice, good and evil and responsibilities of being human.

Ethical philosophy also shows how ethical judgments and ethical statements or attitudes are formed. Ethics is related to self realization about the needs of the human condition – such as doing the right thing at the right time and in the right manner for the right reason is considered virtuous and ethical. Yet the *psychology of ethics* involves more than just understanding moral values and appreciation of the human condition by scientists and engineers.

The *psychology of ethics* is about basic beliefs and attitudes and the formation of these beliefs as also how scientific and engineering value systems are shaped through moral development.

Scientists and engineers often think of psychological reasoning and philosophical reasoning as fuzzy and imprecise. It is true, after all, that the qualitative thinking that is related to the application of ethical principles is not susceptible to the same kind of precision that can be achieved in science or engineering. Often, however, ethical thinking is unnecessarily confused, and much of this confusion is due to the failure to distinguish between three kinds of statements that are made in ethics: (1) factual statements, (2) conceptual statements, and (3) moral statements.

Factual statements (the essence of the scientific and engineering disciplines) are either true or false and refer to claims that can be confirmed or refuted by empirical observation. In discussing factual disagreements, appeal is made to factual or empirical considerations.

Conceptual statements are statements about the meaning or scope of certain terms or principles and discussions of conceptual issues can be very important in ethics. In considering conceptual disagreements, arguments are presented about the appropriateness of one definition as opposed to another.

Moral statements are statements that imply an issue or an action is right or wrong, and, needless to say, there are many disagreements over moral statements. Working on a defense contract to produce the next generation of weapons may allow one scientist to work on the project in good conscience while another scientist could not. In evaluating moral disagreements, appeals are made to broader and more basic moral principles.

It is often recognized that correct actions and behavior involve doing the right thing when it is not in the personal self interest of the scientist or engineer. Sacrifices may have to be made that can never be regained because it is in someone else's interest and because it brings about the greatest good for the greatest number of deserving people.

3.3 Ethics in Science and Engineering

There is a widespread assumption that scientists and engineers conducting basic or applied research should not neglect the fact that their work can ultimately have a great impact on society.

Since the beginning of the seventeenth century, research programs have been used to transform concepts into theories and, simultaneous with this development, there has been some degree of diffusion as researchers have explored new lines of inquiry as they attempt to make their contributions to the literature.

On the other hand, it should be recognized (but not used as an excuse for unethical behavior) that the acceleration of scientific advances in the last few decades raises unprecedented ethical dilemmas. Common moral intuitions are often sufficient for every day moral decision making but such intuitions are insufficient when applied to the problems raised by recent developments in science and engineering. Thus, it seems clear that some specific ethics training is essential to make scientists and engineers mentally and morally equipped to face the emerging scientific and engineering challenges.

However, contemporary ethical concerns do not only cover the potential negative *results* of scientific and engineering activity, but also the *procedures* employed in the laboratory and in the field for conducting scientific and engineering research. There are honest and dishonest ways of doing science. *Falsification of data* and *plagiarism* are the most typical examples of scientific and engineering misconduct. Allegations of scientific and engineering misconduct are not unique to the present time or the last several decades but they now receive a broad coverage in the mass media. In this respect, it is often acknowledged that each time that a new scandal of scientific and engineering dishonesty comes to light, the public trust in science and engineering deteriorates.

The general consensus is that the only effective way to mitigate scientific and engineering misconduct is through education of young researchers. In fact, in context of this book, it is more urgent than ever to provide scientists and engineers with the conceptual tools they need to develop their ethical reasoning.

Every society and professional group has in place a range of norms to guide the behavior of its members. There is a direct correlation between levels of moral outrage expressed and the importance of the norm. In the realm of higher education, norms specify the desired practices with respect to teaching, research and service. Without scientists and engineers in academia, government, and industry would be free to follow their own unconstrained preferences in research.

Thus, norms are indicators of professionalization and also represent what is considered important by a group articulating how professional choices mesh with service. Some have argued however that it is difficult to establish unambiguous ethical standards in academia, and this leads to a range of judgment calls (Whicker and Kronenfeld, 1994). The nature of this challenge is shaped by contextual factors such as societal changes, information overload, competencies all of which impact departmental cultures, individual academic roles and identities. There is a relationship between academic communities and the ideas they express (Becher and Trowler, 2001). Academic culture comprises disciplinary knowledge, growth, enquiry methods, and research outcomes.

Ethical behavior (and, by inference, unethical behavior) in science and engineering is attracting increasing interest in colleges of science and colleges of engineering both on a national and international scale. Evidence of this interest in professional ethics is manifested in the creation of courses in scientific and engineering ethics as a means of introducing ethical issues into required undergraduate science and engineering courses. These courses are increasing in number as students show a willingness to learn more about the philosophical and psychological value of the implications of their actions as professional scientists and/or engineers.

In response to this demand, universities and organizations alike must be prepared to introduce literature and courses on scientific and engineering ethics into their classrooms or education programs.

Increasing levels of subjectivity seem to be associated with improvements in one's qualification in the realm of higher education. As such, professors might display more subjectivity in their conduct than first-time lecturers. High levels of emotion become associated with which students get what, when, where, how and why in terms of grades. The orientation to act unethically is loaded with emotions and not rationality. Of relevance in this regard is underlining of the importance of focusing on values, traditions and collective identities that shape higher education institutions as social entities that are loaded with affect and non-rationality (Gumport, 2007). Academic freedom has to be enjoyed within the constraints of ethical consideration (Chambers, 1983).

As with any other higher-order intellectual activity, resolving moral problems requires that we be both analytical and imaginative. In the analytical mode, a scientist or engineer can recognize the component parts of ethical problems, which assists him to know what kinds of solutions are appropriate. However, resolving ethical problems often requires something more – usually in the form of a philosophical or psychological approach.

Scientists who believe that they deserve more recognition are more likely to falsify, plagiarize or manipulate their data in order to report successful results. Small scale deviant practices continue to be practiced because, despite the canons of scientific and engineering researchers, it is always possible to attribute small inconsistencies to unavoidable errors that accompany or infiltrate all research (Glaser, 1964). One of the major determinants of judgments of the degree of responsibility is whether a controllable act is perceived or intentionally committed or due to negligence (Werner, 1995) but, either way, the outcome is wrong.

Since judgment can only be reliably made after some period if observation or investigation it is essential, however, that the following criteria should be given consideration:

1. whether ethical standards should be known and are clear,
2. whether they are clear but ignored, or
3. whether they are being followed (McDowell, 2000).

However, these criteria should not be used as an excuse to allow the miscreant to be absolved of all blame since he should have applied individual philosophical and ethical principle to the work. Having the three criteria in place can be used to absolve the organization of any blame – the blame must fall squarely on the shoulders of the errant individual scientist or engineer.

Whether or not there is a crisis in the perceived (or real) responsibility of scientists or engineers depends very much on the extent to which individuals were responsible and disciplined before acquiring professional status. The search for truth, knowledge, and understanding of the scientific and engineering world pose strong ethical demands on the scientist or engineer (Guba, 1990). Indeed, methodological, analytical and ethical issues are closely interconnected (Ryen, 2009), particularly so because of the relationships with scientists and engineers involved whose research and whose philosophical and psychological attitudes, values, perceptions of issues vary.

In making judgments about the right conduct or behavior, scientists and engineers must recognize the value of moral consistency. The requirements of consistency take several different forms. An example of moral consistency occurs when a consulting scientist/engineer breaks confidentiality with a client because it is in the scientist/engineer's interest to do so but condemns another scientist/engineer for doing the same thing. The scientist/engineer is not applying the same standards to himself that he expects everyone else to follow. A scientist or engineer must be consistent with his own moral standards. Moral beliefs must be consistent with one another and those relating to confidentiality must be consistent with moral beliefs in professional ethics and any other moral issue.

One way to think consistently in this way is to have a *moral theory*, i.e. a set of moral principles which systematically link moral beliefs to one another by means of a set of coherent moral principles. A moral theory in any area allows the scientist or engineer to have the opportunity to define terms in uniform ways and to relate a set of moral ideas to one another in a consistent manner.

There is always the need for a scientist or engineer to test his moral views for overall consistency. On this basis, it is desirable to have a single moral theory in which all of scientific and engineering views (as they pertain to research) can be included. However, moral philosophers have generally concluded that it is not possible to incorporate all of the moral views that are generally accepted by scientists and engineers into a single coherent moral theory. As a result, there seems to be two systems of moral concepts that are the most influential, although there are considerable areas of overlap: (1) utilitarianism and (2) the ethics of respect for individuals.

The moral standard of *utilitarianism* is: those actions are right that produce the greatest total amount of human well-being. Utilitarianism has intuitive appeal to many scientists and engineers because human well-being seems to be such a natural goal of their respective endeavors. Furthermore, a utilitarian analysis of a moral problem consists of three steps: (1) determination of the audience of the action or policy in question, i.e. those people who will be affected for better or for worse, (2) the positive and negative effects of any alternative actions, and (3) the course of action that produces the greatest overall utility.

Implementation of the utilitarian perspective requires extensive knowledge of facts, which may not be available. This is especially evident in the case of cost/benefit and risk/benefit analysis. In order to balance the cost or negative utility of a scientific or engineering project against the benefit or positive utility, the long-term effects of the project on the public must be calculated. This requires considerable knowledge, some of which may not be available and the long-term positive and negative consequences of an action or policy remain unknown. In such cases, evaluation from the utilitarian perspective involves a *best guess* approach, which may not be very satisfactory.

The moral standard of the *ethics of respect for persons* recognizes equal respect for each human person as a moral agent (an individual capable of both formulating and pursuing purposes of his or her own and of being responsible for the actions taken to fulfill those purposes). Thus, moral agents must be distinguished from things, which exist to fulfill purposes imposed on them by moral agents.

The emphasis on respect for each individual is expressed in the phrase, "do unto others as you would have them do unto you," which requires that a scientist or engineer to consider others by imaginatively placing himself in the position of other members who could be affected by his or her actions. However, this line of thinking or behavior may lead to seemingly perverse results because it seems too permissive and sometimes because it seems too restrictive.

In order to provide a more precise and objective guideline for respecting the moral agency of individuals, some moral philosophers have appealed to a *doctrine of rights*. A *right* is an entitlement to act or to have another individual act in a certain way. *Rights* serve as a protective barrier, shielding individuals from the unjustified infringements of others. The rights necessary to implement the ethics of respect for persons are the rights to freedom and those physical and non-physical conditions necessary to realize the well-being of an individual as preferred by the individual.

Thus ethics in science and engineering is concerned with what ought to be (what we ought to do, what we ought to be, the right and the wrong). Science and engineering, taken very generally, are concerned with *what is* (what the world is like, the true and the false). There is more to science and engineering than a collection of facts. Even if it were possible to know and to express all the truths there are, a complete listing of them would not constitute an adequate scientific or engineering discipline. At a minimum, there is an additional need to subsume particular truths under general laws. Furthermore, a proposed law of science or engineering may cover all the relevant phenomena yet still be unsatisfactory if it lacks explanatory force.

Science and engineering are both attempts to make sense of the world and to explain natural phenomena. In principle, the basic premise of science and engineering is observation. Scientific and engineering theory must account for observations, which are detailed and not simply glances, glimpses, or impressions. Scientific and engineering observations have to measure up to certain standards, which may be more or less well-defined, depending on the scientific or engineering discipline. Scientists and engineers are quite willing to discard supposed observations as simply mistaken, biased, fraudulent, hallucinatory, or otherwise spurious.

A theory constructed to account for a set of observations may end up presenting an explanatory framework that includes most of the observations, but eliminate some. If the theory covers 95% of the observations, but cannot account for the remaining 5%, the deviant 5% are conveniently rejected as being due to *experimental error*, usually an error of some ill-defined or unknown sort. Even though observations are basic, the scientist or engineer is willing to sacrifice observations to theoretical simplicity and/or explanatory power.

Corresponding to scientific observations are *intuitions* of right and wrong, correct and incorrect. Intuitions are not merely transitory emotions or responses just as observations are not glances or momentary impressions. Intuitions are reflective evaluations and a scientist or engineer may at first react to experimental data with disapproval but may, upon brief reflection, reject his reaction and replace it by the intuitive feeling that there is nothing amiss with the data. At a high level of generality the criteria for the adequacy of intuitive reaction are the same as those for the adequacy of scientific theories.

Moral intuitions are significantly affected by a wide range of prior commitments and inclinations, and there are similarities with scientific and engineering observations. Scientific and engineering observations are often *theory-driven* and subject to bias from many sources – the scientist or engineer observes what he is looking for and the experience is categorized within the limits of the conceptual frameworks that are brought to laboratory or to the field. The danger is that the scientist observes what he is looking for, and he observes what he wants to see. Personal commitments, inclinations, and theories influence scientific and engineering observations and our moral intuitions.

Progress in both science and engineering is a matter of developing theories of increasing inclusiveness and coherence, which make sense of our intuitions and discipline them.

In this respect, academic institutions play an important role in ensuring that scientific and engineering students develop better moral reasoning skills and that they learn to behave ethically from the start of their professional lives. Although this need is widely recognized in theory, it remains largely neglected in practice. Young scientists and engineering have a real need for specific ethics training and for more opportunities to discuss about the ethical dimension of their work (Andorno, 2003).

In fact, belonging to a group (such as a society or association of scientists and engineers) means following basic standards of conformity and conformity determines the extent to which social behavior would be in accordance with what the Society or Association accepts or considers as standard (Chapter 6). Standard behavior is, in fact, closely related to ethical behavior, which about conformity of behavior and doing what is right according to social standards or values.

For example, when considering developmental psychology, individual needs are met through social conformity as following ethical standards and engaging in ethical behavior, which can be rewarding to an individual and which would encourage or reinforce ethical standards. Ethics fulfils our social and recognition needs and our moral needs of regulating the desires of scientists and engineers.

Ethics can be considered as the moral aspect of the psychic structure of scientists and engineers and is essential to group behavior. Ethics is an important social developmental process and ethical values and beliefs must be forged early in the careers (or lives though parental guidance or educational system) of scientist and engineers (Chapter 5).

3.4 A Phenomenological Theory of Ethics

Phenomenology is a philosophy or method of inquiry based on the premise that reality consists of objects and events as they are perceived or understood in human consciousness and not of anything independent of human consciousness.

Human nature is inherently self-centered. Emotional expressions (mild to extreme) dominate human behavior. While such emotionality is natural, thinking is a process that requires time and has to be cultivated. As such, the process for becoming increasingly rational or ethical is also time consuming and its outcomes are not guaranteed to evolve and emerge as expected. Emotion, however, remains at the core, intensifies and establishes an inverse relationship with the development of rational or logical thinking.

The continued presence of emotionality is necessary to feel desire, passions. Emotions are the flames that will fire up our rational thinking in a manner that will move us to act on our thoughts. Without emotions, the completely rational person would remain unmoved. Without the acquired ability to reason, the completely emotional individual would be unrestrained to get whatever he or she wants. The baser or basic instinct of humans would definitely run amok since our underdeveloped reasoning skills would not be able to constrain or slow down the emotional onslaught.

The willingness and orientation to conform to ethical standards and make proper ethical judgments is rooted in the extent of internalization of a sense of morality in earlier years of one's socialization. The socialization of reasoning lags behind the emotional development of the individual and, thus, moral decision-making is guided by greater levels of emotionality rather than rationality. As such, moral judgments are more subjective and the ability of a scientist or engineer to reason about ethical issues in later years is affected by the extent to which one has emerged out of the emotionality of the past Self. Indeed, given this situation, some individuals may not subscribe to the notion of ethical fact and their ethical judgments would not really be as objective (Hurley, 2007).

Phenomenology posits that: (1) scientists and engineers live in a natural world of values, assumptions, judgments, feelings and free choices, and (2) subjectivity and objectivity are intertwined. This phenomenological perspective on ethics emphasizes the role of the individual differences in meanings as providing the impetus for chosen actions.

Objectivity and subjectivity correspond to cognition and emotions respectively. With respect to ethical/unethical practices, academia is one of several contexts in which such scenarios happen. In all such sittings, the authors posit that individuals take aspects of social life (i.e. those that they were socialized into) with them and these aspects interact and are assessed by the individual's subjectivity and objectivity. For those individuals who are more emotionally driven (subjective) their ultimate actions would reflect their feelings. For those individuals who reflect on consequences in a rational and objective manner, their final actions might reflect the same. In situations where individuals know better and act otherwise, this may be due to emotionally loaded cognitions or outcomes of emotions (Phelps, 2005).

Increases in the level, frequency and severity of ethical dilemmas are directly linked to the quality of humanizing experiences, consideration for self, other sources of influence and the reward system of a society. Questionable socialization practices, whether self-inflicted or not, usually orient people in the long term to develop certain beliefs, values, and attitudes. These beliefs, values and norms can influence people independent of socially acceptable standards. Such individuals would experience a greater propensity to deviate from normative behavior in any context. This is the basis for unethical, inconsiderate conduct. For such individuals the extent to which their literature reviews, methodology, data analysis, and discussion of findings are indefensible is of little concern to them. Of greater importance is their ability to make "wrong" look "right" in order to convince others and be rewarded.

The proliferation of unethical practices is both a culmination and constituent of *values in conflict* which leads to a focus on ends instead of process. Ethics and morals both equally refer to prescriptive rules or principles of action, rules presumably designed to make things better in some important sense by guiding appropriate action (Wagner and Simpson, 2009). In later years the word, "ethics," was used only in association with rules and regulations to govern the actions of professionals and morals were used only in reference to personal conduct.

There is a concern that both the social and personal questionable sense of morality at the individual level is evidence of the weakening of sociality, the loss of a sense of collegiality and the emergence of individualism even in the tertiary education environment. Differences in moral, personal and interpersonal orientations are due to a division between individualistic and collectivistic, independent and interdependent, bounded and unbounded (Turiel, 2002).

Moral identity formation is part of personality development. "The more you identify with a moral standard…the more that standard will direct your attention and flavor and filter human perception and interpretation. Aspects of the moral experience across space and time include being conscious of different interpretations of reality, different emotional expressions and exposure to differential degrees of commitment to fairness" (Peterson and Seligman, 2004).

"Ethical questions are essentially linked with choice" (Bali, 1997). Depending on the choice that an individual makes, there are certain consequences in the form of positive, negative, indifferent or delayed actions and reactions from others in the short or long term. Such actions on the part of subjects or objects may or may not conform to proper ethical conduct. Any judgment of the ethics of the action is however dependent on organizational and societal requirements as embedded in Codes of Ethics (Chapter 6). In democratic scientific and engineering societies (Chapter 5) there is an expectation that individuals would align their behavior with that of others in order to promote general well-being.

There is a big difference between personal goals and public morality (Hakim, 2003) and, as a result, individuals are willing to engage in unethical practices particularly if they believe that such would not be detected and/or because of prioritization of the personal in the context of the social. The greatest challenge then is how to balance individuality with sociality in any context. This is further aggravated in circumstances where one enjoys freedom of choice in a context of deteriorating social norms and unenforced rules, regulations and sanctions. In pursuit of moral standards or proper ethical conduct, one set of standards must be capable of getting general agreement from all who can think rationally (Bali, 1997).

Ethics is indeed one of the pillars of scientific research, teaching and community service, thus one of the requirements of higher education. It is definitely one of the criteria for evaluating the quality of higher education in these aforementioned areas. Despite the range of factors that contribute to (un)ethical behavior, the central determinant is personal norm which determines the meaning that an individual scientist or engineer attaches to his position with respect to ethics. Personal norms can override the influence if any other factor including the codes of ethics of professional bodies. The ability of a scientist or engineer to manage emotions during the process of coming to know orients many individuals to act on feelings and engage in unethical practices.

This is followed by further discussion and clarification of meanings which results in less emotionally loaded arguments and more rational views. At this stage there might be further reflection and research on the outcomes of the discussions and depending on this greater acceptance of the views advanced.

3.5 Conflicts of Interest

All members of the scientific community are faced with balancing conflicting interests and there is growing concern by many that a commitment to profit has resulted in a loss of confidence in the integrity of institutions of higher education and scientific and engineering research.

A *conflict of interest* occurs when a scientist or engineer or their respective organization becomes involved in multiple interests, one of which could possibly corrupt or destroy the integrity of the other.

A conflict arises when a scientist or engineer has the potential to lose impartiality because of the possibility of a clash between the scientist or engineer's self-interest and professional interest or public interest. Thus, it is a situation where a scientist or engineer limits his ability to discharge duties responsibility to another party (be it a scientist or engineer or the public).

A conflict of interest also exists if a scientist or engineer is entrusted with some impartiality. A modicum of trust is necessary to create it. The presence of a conflict of interest is independent from the execution of impropriety and can be discovered and voluntarily negated before any corruption occurs.

Professional ethics need to be set, especially in a business or a large organization where a group of individuals (in this context, scientist and/or engineers) may find themselves in situations where their respective values are in conflict with another, and they are in need of some reference as to what is considered ethical and not.

Briefly, a conflict of interest may be described as a situation where your personal interests or activities could influence an individual's judgment or decision-making and consequently, your ability to act in the best interests of the company or business.

Generally, there are three categories of conflict situations:

1. real conflicts of interest,
2. situations that constitute potential conflicts, and
3. situations that are likely to be perceived as conflicts of interest.

Each type of conflict presents its own problems for scientists and engineers and requires careful review and appropriate management or elimination. They all have one thing in common: Unless they are addressed adequately, they will cause a loss of public trust in the institution and the scientific or engineering research it conducts. Over time, an institution (and it scientist and engineers) that has a reputation for being indifferent to conflicts of interest can suffer a number of consequences: loss of prestige, a lessening of respect for the scientists and engineers or faculty, and suspicion that research findings are tainted and/or manipulated.

On the other hand, a *conflict of commitment* (often interwoven with, and related to, a conflict of interest) is, generally, a situation in which a scientist or engineer is dedicating time to personal activities in excess of the time permitted by institutional policy, or to other activities that may detract from his or her primary responsibility to the institution. The issue is whether the scientist or engineer's s commitment of time and effort are inconsistent with commitment to the organization/institution and its interests.

Some examples of conflicts of commitment (taken from academia since the corresponding behavior patterns from a company are not known) are: a faculty member dedicating more than the permitted one day per week on personal consulting with a company or companies, a faculty member accepting an unpaid position on a company's scientific board of advisors and having access to and/or divulging confidential information when the company is sponsoring the faculty member's research, or a faculty member uses institution resources, including office or laboratory space and secretarial services in support of his personal consulting.

Again, using academia as the example, a conflict of commitment also exists when a scientist or engineer faculty member has instructional and mentoring responsibilities and he uses graduate students on a personal consulting project. While graduate students may be interested in the work performed on the consulting project, their participation is primarily of personal benefit to the faculty member. It is a misuse of a graduate student's time and detracts from his or her efforts to complete degree requirements. Furthermore, due to the intellectual property and confidentiality provisions included in most consulting agreements, graduate students would be unable to publish the results of their work.

A code of ethics enables the organization/institution to establish the ideals and responsibilities of the profession or business and also serves as a reference on acceptable conduct, increases awareness and maintains consistency and ensures improved quality. When scientists and engineers follow a set code of conduct, it also enables their colleagues and clients customers to trust the scientist or engineer with their critical information and is a conscious effort to protect the interests of the clients and professionals. This is especially true when a scientist or engineer is in business as a consultant. The client must know and feel comfortable that any facts or date disclosed to the consultant will not appear in a report or presentation to another client or as an additional slide in a presentation at a scientific or engineering symposium.

Thus, scientists and engineers must exercise reasonable care and sound judgment to achieve and maintain independence and objectivity in their business-related activities. They must not intentionally conceal or misrepresent information or facts relating to recommendations, actions and findings or reveal any kind of information to deceive their customers, clients or partners, as the case may be. In short, the philosophical and psychological aspects of being a professional scientist or engineer involve doing the right things in the interest of the organization, profession or business, as the case may be.

On this basis, it is assumed that professionals maintain confidentiality and do not disclose sensitive and critical information about their clients/customers to third parties. The exceptions to this statement are if the scientist or engineer is required to disclose by law, if the information concerns illegal activities on the part of the client, or if the client expressly permits disclosure of information.

Conflicts between competing obligations, both of which appear to be valid, are common features of scientific and engineering life. The issue is the means by which such conflicts can be approached.

The most obvious criterion for separating conflicts (i.e., separating legitimate activities or obligations of scientists and engineers employees from illegitimate activities or obligations) is that the scientist/engineer employee has an obligation to avoid any activity that interferes in a clear and direct way with ethical performance.

If a potential conflict does arise, the scientists/engineer should make his feeling obvious in as responsible a manner as possible. If a corporate hierarchy is involved, this should be done in a private and non-confrontational way. The scientist/engineer should do everything possible to avoid embarrassing the employer and give the employer the opportunity to correct the problem.

Thus, if disclosures are submitted indicating that a real or potential conflict exists, the organization/institution must have in place (if not, develop quickly) a plan to eliminate or mitigate the conflict. The plan must also contain monitoring provisions so that the organization/institution can be assured that the plan is successfully implemented.

Certain scientific and engineering professionals (especially those who may appear as an expert witness in court) are required either by legal rules or by rules related to the respective professional organization, or by society statute or association statute to disclose any actual or potential conflicts of interest. In some instances, the failure to provide full disclosure is a crime. Any scientists or engineers with a conflict of interest are expected to recuse themselves from (i.e., abstain from) decisions where such a conflict exists. This is in keeping with common sense ethics, codified ethics, or by (as noted above) by legal statute. In fact, to minimize any conflict, the scientists or engineer should not participate in any way in the decision, including general or specific discussions on the particular issue.

Generally, a code of ethics will forbid conflicts of interests. Often, however, the specifics can be controversial. For example, in the current context, it is arguable (often convincingly so) that a professor should not be allowed to have extra-curricular relationship or an extra-professional relationship with a student. There are those who will argue that the outcome should depend on whether the student is in a class of, or being advised by, the faculty member. It must also be remembered that the professor may sit on a committee (some levels removed from the usual professor-student level) that eventually decides whether or not the student has sufficient abilities to graduate. Is this not a conflict?

Some will argue "no" on the basis that the professor (no matter how eloquent his argument) is outnumbered. The same is true of many industrial settings in regard to a relationship between a supervisor (at whatever level) and an employee (at a lower level) in the company.

In-place codes of ethics in any organization can help to minimize problems with conflicts of interests before such conflict arise. The codes of ethics, assuming that they give details, can state the extent to which such conflicts should be avoided, and state what the involved parties should do where such conflicts are permitted by a code of ethics. In such cases, scientists and engineers cannot claim that they were unaware that their improper behavior was unethical. It is also important to recognize that that the threat of disciplinary action (whatever is prescribed by the code of ethics) helps to minimize unacceptable conflicts or improper acts when a conflict is unavoidable.

Disclosure of potential conflicts of interest is a key issue because the organization/institution must review each case and make a determination as to whether a conflict exists or not. To accomplish this, many organizations/institutions have formed a conflict of interest committee. Scientists and engineers (no matter how logical their argument might be) cannot make this decision for themselves. It must be done by those having (in this context) no scientific or engineering or financial interest at stake. It must also be done by those appointed by and representing the organization/institution.

Conflict of interest committees have a range of options available to them when a conflict of interest needs to be managed. Each situation will have its own unique aspects and, therefore, the committee must carefully review each situation and ensure that the problems or potential problems are adequately addressed. At the same time, it is important that the committee not micro-manage the situation so that the scientific and engineering research is not impeded unnecessarily. Some of the options available to the conflict of interest committee are:

1. require that the researcher disclose his or her conflicting interests to all collaborators;
2. require that the researcher have a person unaffiliated with his or her research provide an objective review of any manuscripts intended for publication;
3. require that the researcher revise aspects of the research project so as to mitigate any real or potential conflicts; and
4. prohibit the researcher from participating in certain proposed research activities.

However, while institutions can create policies, criteria, and guidelines, each case has its unique characteristics and must be evaluated on its own merits. Sometimes a case may present very obvious problems that must be managed. Sometimes two cases may be very similar and yet managed in very different ways. The bottom line is that each case must be reviewed in light of the policies of both the institution and the scientist and engineers involved.

Finally, conflicts of interest are not inherently bad or unethical. It is the failure to acknowledge and report real and potential conflicts or the failure to manage them effectively that are clear violations of the relevant policies.

In summary, the methodology of choice is that whenever conflicts of interest interfere with the conduct of scientific and engineering research, the work should not be undertaken (Bok, 2005). The best way to handle conflicts of interests is to avoid them entirely.

References

Altman, E. 1997. "Scientific Research Misconduct." In *Research Misconduct: Issues, Implications and Strategies*. E. Altman and

P. Hernon (Editors). Greenwich: Ablex Publishing, New York. Page 22.

Andorno, R. 2003. "Teaching Ethics to Scientists: An Experience Report." *Ethik in der Medizin*, 15(3): 243–245.

Audi, R. 2009. *Business Ethics and Ethical Business*. Oxford University Press, Oxford, United Kingdom.

Becher, A., and Trowler, P.R. 2001. *Academic Tribes and Territories. 2nd Edition*. ed. Buckingham: Open University Press, McGraw-Hill Education, Maidenhead, Berkshire, United Kingdom.

Bok, D. 2006. Universities *in the Marketplace*. University of Princeton, Princeton, New Jersey.

Brown, B.L. and Hedges, D. 2009. "Use and Misuse of Quantitative Methods." In *The Handbook of Social Research Ethics*. D.M. Mertens and P.E. Ginsberg (Editors). Sage Publications, Thousand Oaks, California. Page 373–390.

Brydon, L. 2006. "Ethical Practices in Doing Development Research." In *Doing Development Research* edited by V. Desai an R.B. Potter. Sage Publications, Thousand Oaks, California. Page 25–33.

Chambers, C. 1983. "The Social Contract Nature of Academic Freedom." In *Ethical Principles Practices and Problems in Higher Education*. M.C. Baca and R.H. Stein (Editors). Charles Thomas Publishers, Springfield, Illinois. Page 23–36.

Crane, A., and Matten, D. 2004. *Business Ethics*. Oxford University Press, Oxford, United Kingdom.

Glaser, B.G. 1964. *Organizational Scientists : Their Professional Careers*. Bobbs-Merrill Company, Indianapolis, Indiana.

Gumport, P.J. 2007. "Sociology of Higher Education: An Evolving Field." In *Sociology of Higher Education: Contributions and Their Contexts*. P.J. Gumport (Editor). Johns Hopkins University Press, Baltimore, Maryland. Page 17–50.

Hakim, C. 2003. "Public Morality Versus Personal Choice: The Failure of Social Attitude Surveys." *British Journal of Sociology*, 54(3): 339–345.

Hurley, P. 2007. "Desire, Judgment and Reason: Exploring the Path Not Taken." *The Journal of Ethics*, 11(4): 377–403.

Mark, M.M., and Gamble, C. 2009. "Experiments, Quasi-Experiments and Ethics." In *The Handbook of Social Research Ethics*. D.M. Mertens and P.E. Ginsberg (Editors). Sage Publications, Thousand Oaks, California. Page 198–213.

McDowell, B. 2000. Ethics *and Excuses: The Crisis in Professional Responsibility*. Quorum Books, Westport, Connecticut.

Peterson, C., and Seligman, M.E.R. 2004. *Character Strengths and Virtues*. Oxford University Press, Oxford, United Kingdom.

Phelps, E.A. "The Interaction of Emotion and Cognition: Insights from Studies of the Human Amygdala." In *Emotion and Consciousness*.

L.F. Barrett, P.M. Niedenthal and P. Winkelman (Editors). The Guilford Press, New York. Page 51–66.

Shrader-Frechette, K. 1994. *Ethics of Scientific Research*. Rowman and Littlefield Publishers, Inc., New York.

Turiel, E. 2002. *The Culture of Morality: Social Development, Context and Conflict*. Cambridge University Press, Cambridge, United Kingdom.

Viswanathan, M. 2005. *Measurement Error and Research Design*. Sage Publications, Thousand Oaks, California.

Wagner, P.A. and Simpson, D.J. 2009. *Ethical Decision Making in School Administration: Leadership as Moral Architecture*. Sage Publications, Thousand Oaks, California.

Whicker, M.L., and Kronenfeld, J.J. 1994. *Dealing with Ethical Dilemmas on Campus*. Sage Publications, Thousand Oaks, California.

Wolf, A., Turner, D., and Toms, K. 2009. "Ethical Perspectives in Program Evaluation." In *The Handbook of Social Research Ethics*. D.M. Mertens and P.E. Ginsberg (Editors). Sage Publications, Thousand Oaks, California. Page 170–183.

Yow, V.R. 2005. *Recording Oral History*. Rowman and Littlefield Publishers, New York.

4

Education of Scientists and Engineers

4.1 Introduction

Education is basic to achieving personal and national goals in science and engineering. Universities lay the groundwork for the scientists and engineers of tomorrow, but, in practice, this is not always the case as many universities do not make any great effort to teach the practical aspects of the various science and engineering courses (ACCN, 2003). The universities educate, to a point, future teachers and researchers but the extent of this education is often insufficient for life in the non-academic world. However, graduate scholarship and research are (supposedly) key contributors to meeting broad national goals of technological, economic, and socio-economic development (Knight, 2002).

Yet, academic cheating is a common phenomenon in middle schools, high schools, and colleges (Cizek, 1999; Evans and Craig, 1990a, 1990b; Leveque and Walker, 1970; Schab, 1991). In 1987, the California Department of

Educational groups labeled cheating an epidemic after finding that 75% of secondary school students reported that they had at some time cheated on school work (Schab, 1991). In addition, from the early 1960s through the 1990s, cheating among students increased (Baird, 1980; Schab, 1991), as if accompanying the tendency for parents to litigate against the school or university when a student is reprimanded for any unsocial or illegal activity.

Indeed, university students often report that they cheated more in high school than in college (Baird, 1980). There is also evidence that cheating is more widespread in high school than during middle school (Brandes, 1986; Evans and Craig, 1990b). This is unacceptable at any level of education, whether it be at the pre-high school level, the high school level, or the university level.

Persons educated as scientists and engineers are meant to provide service to society via their development of original ideas, which are brought to fruition in teaching, industry, business, and government. Graduate students often go beyond the thinking of their professors and create a new generation of science and engineering thought. The student learns from the professor, but the professor, if he will admit it, also learns from the student.

By educating students in the context of research, the systems for the education of scientists and engineers have set national standards for preparing scientists and engineers for research careers in academia, government, and industry. Furthermore, by attracting outstanding students and faculty members (hopefully who have some understanding of the non-academic world) national systems have, to some extent, benefited from an infusion of both talent and ideas.

In spite of the examples presented below, some students do not cheat. Students, high in measures of honesty, report that they would feel an extreme level of guilt if they cheated. There appears to be a general consensus between students and teachers/professors as to reasons for ignoring cheating, which suggests a general impression of denial (Staats et al., 2009). This attitude of non-reporting is, in itself, a form of cheating.

It is the responsibility of the teacher/professor to confront cheating and misconduct head-on. It is these people, who are in the position of responsibility, who should recognize cheating and misconduct, and who must offer a suitable punishment as a deterrent.

4.2 The High School Experience

Evidence indicates that cheating is very common during adolescence (Anderman et al., 1998). A number of studies indicate that middle school environments are more focused on grades and performance than are elementary schools (Midgley and Urdan, 1995; Anderman and Midgley, 1997). As students move from elementary schools into middle schools, the increased focus on grades may lead some students to cheat (Anderman et al., 1998). In addition, the data also indicates that cheating was most prevalent among white males attending private schools. Furthermore, students who felt alienated from school were more likely to cheat (Calabrese and Cochran (1990).

On this basis, by the time the student reaches high school the tendency for cheating and misconduct may be a set pattern. The issue is related to the nature of the punishment, if any, for the student proven guilty of cheating. If there is no form of punishment or punishment is very light, such as a slap on the wrist, (and the records are usually dim or non-committal on this subject) the student enters or continues in high school knowing that he can progress through the school system using whatever means he choose. Fair means or foul.

Many students start thinking about the possibility of a career when their interest is ignited by a high-school or undergraduate teacher or some other role model (Sadler et al., 2006). This is the time to start meeting and talking with scientists and engineers in fields of interest. Science and engineering fields are prestigious fields of education and as they bring greater financial rewards to the graduates than arts and humanities and even the social sciences. These early contacts can be crucial in helping the students to navigate the terrain of science and engineering as they move through their career.

The standards of education require that students increasingly engage in inquiry-based, collaborative learning experiences that emphasize observation, collection, and analysis of data from student-oriented experiments. They also stress the importance of helping students learn about the relationships among the sciences and the relevance of science, mathematics, and technology to other realms of inquiry and practice.

Students from the United States have demonstrated a steady decline from the 4th through the 12th grade in their mathematics and science performance. By the 12th grade, American students rank near the bottom of every category for knowledge of both general and advanced levels of science and mathematics, compared with their counterparts in countries around the world.

Students who do arrive at college with what traditionally has been considered good preparation in science and mathematics may not have actually developed a real conceptual understanding or the ability to solve problems; particularly in mathematics and the physical sciences, when compared with students in other countries with similar educational backgrounds. Yet, at the present, in the United States, students with high exam scores in hand can sometimes avoid taking any further science or mathematics classes at the postsecondary level; therefore, allowing them to think about the subject matter more deeply.

If the mission of post-secondary education is to provide students with opportunities to experience and think about subject matter more deeply than they could in high school then allowing some students to complete or waive specific graduation requirements on the basis of high examination scores alone could be self-defeating to that mission, compared to awarding them credit toward the total number of credits required for graduation.

Requiring all students to complete introductory, interdisciplinary, or higher level courses, regardless of their intended major, would enable some of the best students in the university to experience and appreciate the wealth and breadth of the sciences that they otherwise might have missed during their high school years. In collaboration with the University Office of Admissions, departments should make clear how the various departments will regard students with high scores on these examinations; especially those who wish to use these scores to avoid taking college-level mathematics or science courses.

In the near future, students who have had a standards-based education at the pre-college level, where they engaged in inquiry-based, collaborative learning experiences, will expect to receive more of the same in their undergraduate science and mathematics courses. Postsecondary institutions that take the lead in offering undergraduate curricula of high value to all of their students not only will have highly successful graduates but also will attract the highest quality incoming students.

However, by the time the student enters the university system the trend for misconduct may have been initiated and even set. In spite of the care and attention supposedly showered on students in school, a recent survey shows that cheating in school continues to be rampant. In a survey of 30,000 students across the United States (Josephson Institute, 2009), the results showed that young people are almost unanimous in saying that ethics and character are important on both a personal level and in business, but they express very cynical attitudes about whether a person can be ethical and succeed. Moreover, an alarming number of students admitted to recently lying, cheating, or stealing.

A substantial majority, 64 percent, cheated on a test during a high school year, while 38 percent did so two or more times, up from 60 percent and 35 percent, respectively, in 2006. Students attending non-religious independent schools reported the lowest cheating rate (47 percent) while 63 percent of students from religious schools cheated. Responses about cheating show some geographic disparity: Seventy percent of the students residing in the southeastern United States admitted to cheating, compared to 64 percent in the west, 63 percent in the northeast, and 59 percent in the midwest. More than one in three (36 percent) said they used the Internet to plagiarize an assignment. In 2006 the figure was 33 percent (Josephson Institute, 2009).

In addition, more than one in four (26 percent) of high school students confessed they lied on at least one or two questions on the survey, and dishonesty on surveys usually is an attempt to conceal misconduct. Furthermore, despite these high levels of dishonesty, the respondents have a high self-image when it comes to ethics, 93 percent said they were, "satisfied with their personal ethics and character," and 77 percent noted that, "when it comes to doing what is right, I am better than most people I know" (Josephson Institute, 2009).

It can be argued that there are a variety of reasons why young people still engage in unauthorized use of published material, but one of them is that they just do not believe that copyright laws are morally justified. Instead, they see the benefit and the opportunities presented in easier sharing and distribution of works.

Such thinking is just an easy way out of hard work. Cutting and pasting material from the Internet or from a work by someone else is much easier than having to think through the problem and use individual thoughts to reach meaningful conclusions.

The social cost of such activities is high and the pathway is made for cheating in adult life, through the baccalaureate experience and various advances degrees to work as a professional (Carpenter et al., 2004). There are also implications for the next generation of scientist and engineers.

In a society that is currently (and may always have been) saturated with cynicism, it is easy for young people to justify faulting copyright laws and other forms of misconduct on the basis that everyone else does it – although it does seem that females students are less likely to engage in misconduct than male students (Becker and Ulstad, 2007). Perhaps these are the same young people who criticize the older generation for laxity and the onset of global climate change.

As a final note on the issue of cheating and misconduct at the high school level, cheating on an assignment and/or during an examination hinders learning. While many students cheat on assignments their actions are not looked upon as cheating but as means of learning (Kohn, 2007). Surely this is a meager excuse and only a poor means of justifying cheating. On the other hand, cheating in an examination may also be considered a means to an end that will cease once the student is entrenched in the university system – but this is not the case.

4.3 The Baccalaureate Experience

The baccalaureate degree is the bachelor's degree and carries a, designation related to the broad subject area such as BSc (Bachelor of Science), BEng (Bachelor of Engineering), and so on. In many European countries, the majority of Bachelor's degrees are now honors degrees. Until the mid-20th century, some candidates would take an *ordinary* degree, and then be selected to go on for a final year for the *honors* degree.

A first degree course is usually three years, but it might be reduced to two either by direct second year entry (for people who have done foundation degrees or changed subject) or by doing compressed courses (which are being piloted by several newer universities).

Ordinary degrees are unclassified degrees awarded to all students who have completed the course and obtained sufficient marks to pass the final assessments and examinations. Although ordinary degree courses are often considered to be easier than honors degree courses, this is not always the case, and much depends on the university attended and the subject being studied. Some modern universities offer the opportunity for ordinary degree students to transfer to an honors degree course in the same subject if an acceptable standard is reached after the first or second year of study.

Honors degrees are of a superior academic standard and is awarded in one of four classes depending upon the marks gained in the final assessments and examinations. The top students are awarded a *first class degree*, the next best, *an upper second class degree* (usually referred to as a 2:1), the next a *lower second class degree* (usually referred to as a 2:2), and those with the lowest marks gain a *third class degree*. An *ordinary* or *unclassified* degree (which does not give the graduate the right to add *(Hons)* may be awarded if a student has completed the full honors degree course but has not obtained the total required passes sufficient to merit a third-class honors degree. Alternatively a student may be denied honors if he has had to retake courses.

Many universities in the United States award bachelor's degrees with Latin notifiers, usually (in ascending order): *cum laude* (with honor/praise), *magna cum laude* (with great honor/praise), *summa cum laude* (with highest honor/praise), and the occasionally seen *maxima cum laude* (with maximal honor/praise). Requirements for such notations of honors generally include minimum grade point averages

(GPA), with the highest average required for the *summa* distinction (or *maxima*, when that distinction is present).

Baccalaureate degrees in the United States are typically designed to be completed in four years of full-time study, although some programs (such as engineering or architecture) usually take five, and some universities allow ambitious students (usually with the help of summer school) to complete them in as little as three years. Some universities have a separate academic track known as an *honors* or *scholars* program, which is generally offered to the top percentile of students (based on the grade point average), that offers more challenging courses, more individually-directed seminars, or research projects in place of lieu of the standard curriculum. The students are awarded the same bachelor's degree as students completing the standard curriculum, but with the notation *in cursu honorum* on the degree certificate. Usually, the above Latin honors are separate from the notation for this honors course, but a student in the honors course generally must maintain grades worthy of at least the *cum laude* notation.

If the student has completed the requirements for an honors degree only in a particular discipline (e.g., Chemistry), the degree is designated accordingly (e.g., BSc with Honors in Chemistry). In this case, the degree candidate will complete the normal curriculum for all subjects except the selected discipline. The requirements in either case usually require completion of particular honors seminars, independent research at a level higher than usually required (often with greater personal supervision by faculty than usual), and a written honors thesis in the major subject.

Advances in technology and expanding roles create complex ethical and moral dilemmas for many professionals, especially scientist and engineers (Martin et al., 2003). Having come through a high school system where unethical behavior may be rampant, modification of personal values and of professional scientific and engineering values are both important parts of ethics development in university education. Course content related to essential moral and ethical dilemmas is not routinely included in formal scientific and engineering curricula, but may be taught informally through unplanned discussions with various professors and, even then, individual professorial insights into ethical and/or unethical behavior would vary.

Teaching or instilling values of research ethics ideally falls at the department level and some universities offer a stand-alone course for students to meet federal standards for grant (funding) applications. However, it is often the case that no one assists the students and trainees to understand the subtler and more common ethical problems, especially when the focus of the course is the extreme behavior that counts as research misconduct (Ritter, 2001).

The academic culture of a department is a combination of intended and unintended outcomes that emerge from each of the facets of departmental organization. The nature of the ethical environment depends on how this impacts at the individual level. Academic leaders are basically mandated to enforce policies, rules and regulations. The manner in which that is done depends on the administrative style of academic leaders.

Despite the various theories of management, management in practice can be conceptualized as: (1) operational, focusing, and monitoring of day-to-day activities, and (2) coordinating objectives, strategies, resources and evaluation in a manner that would enhance individual-level competencies and performance. In the absence of such a focus, individual and departmental level activities can become caught in a quagmire of routine activities that may/may not be punctuated by individual-level achievements amidst deteriorating levels of staff morale.

In another scenario, communication and decision making procedures are important. For both of these functions, the formal and informal dimensions must be woven together.

Any attempt to peripheralize one dimension would provide space for a simultaneous: (1) resistance and a hardening of views, and (2) space for the other dimension to blossom.

The values of the scientific and engineering professions are found in their relevant professional code of ethics, but students at the baccalaureate level are not introduced to such codes of ethics until they graduate and decide to become members of their respective societies.

The universities, colleges, and engineering institutes in the United States enroll a larger proportion of young adults than in any other nation. Half of these students will eventually receive bachelor's degrees. For the last three decades, approximately one third percent of bachelor's degree recipients (that is, about 15 percent of each high school graduating class) received their degrees in science or engineering. In recent years, about one-tenth of these bachelor's-level scientists and engineers have gone on to earn science or engineering doctorates. But the question is, outside of religious institutions, "how many of these students have received any formal training in ethics, honesty, and moral behavior as applied to the practice of their scientific and engineering disciplines?"

Many baccalaureate students lack any form of competency in ethical decision making, as it is not an identified expectation of the baccalaureate degree graduate. Values, both personal and professional, do not provide a systematic foundation for ethical decision making. An understanding of ethical principles and theories as well as application of them to the role of the professional scientist and engineer is lacking. The simplest format for presentation of such a course could include ethical theories and principles and their application to the practice of science and engineering.

Placement of separate required ethics courses remains an issue because of the overwhelming amount of content in baccalaureate degree curricula. If ethics content is integrated throughout the curriculum, it should be presented early with continual reinforcement and with the use of a specific ethics textbook. Research indicates that students who have completed an ethics course not only know the correct ethical action but are more likely to implement it. In fact, an understanding of ethical principles and theories as well as application of them to the role of the professional scientist and engineer is essential to ethical decision making in professional practice (Gaul, 1989).

Although 30 percent of baccalaureates are awarded in science and engineering, the relative popularity of different fields has shifted substantially with events in the job market of the last three decades. Increasing college enrollments meant more science and engineering baccalaureate recipients. In contrast, the proportion continuing on to PhD study reflects market demand, the availability of research and development funds, and direct student support.

Most fields of graduate study in the sciences, as distinguished from engineering, are oriented toward the academic as well as the industrial job market; somewhat less than half of PhD scientists work in academic institutions. The PhD is the basic professional degree in most fields of science, and most science students seek research or teaching positions. Despite growing undergraduate enrollments from the late 1960s to the present, a stagnant academic job market and slower growth in federal and commercial research funds have left many young PhD awardees underutilized.

Full-time graduate enrollments in science and engineering have grown in the past decades at many universities, but if not for the influx of foreign graduate students these enrollment increases would have been substantially less. Retirements and turnover of faculty in the mid-1990s, combined with resurgence in undergraduate enrollments later in the decade, offered some relief to these pressures. As a result, the attractiveness of an academic career is still the path forward for many students.

In addition, many institutions, beset by a tenured faculty, continue to hire faculty, even in difficult economic times, which turns into a job for life and involves teaching only a few hours a week and routinely take off an entire year in the name of sabbatical leave while being paid $100,000 or more per year (Shea, 2010).

In engineering and some fields of science, the bachelor's or, increasingly, the master's degree is the most important professional degree. The employment markets for these fields are dominated by industry rather than academia. Because their periods of training are shorter, enrolled students can react more quickly to employment opportunities. These fields, not coincidentally, have been the ones that experience enrollment and employment booms, and subsequent downturns. When there is a downturn, faculty shortages develop. However, foreign faculty have proven vital to maintaining teaching capacity in these fields since US citizens have generally seek high-paying baccalaureate-level industrial employment, rather than graduate study in pursuit of faculty positions.

The undergraduate-baccalaureate years are probably the best chance for the student to take a broad variety of classes outside the primary discipline that might be useful later; such classes could include subject matter related to ethics, honesty and moral behavior during the practice of the profession be it in industry or academia. In addition, classes in sociology, history, philosophy, English (with emphasis on composition), foreign language, and psychology, spread through the undergraduate years are useful in helping a student to acquire understanding, different experiences, and professional maturity.

Students are inundated with ethical questions and choices, the most frequent are whether or not they should behave ethically during testing, participate in unauthorized group homework, and/or plagiarize from the Internet (Szabo and Underwood, 2004). Many factors influence students' decision making processes. Variables related to cognitive development and environment affect how they choose to behave (Bandura, 1991; Love and Simmons, 1998).

An effective way for students to learn about undergraduate education in ethical issues is to join (or form) a study group to share concerns (Nadelson, 2007). In a university setting, the student will meet with undergraduate students, graduate students, and postdoctoral researchers and gain valuable insights. The students can also join student chapters of scientific and engineering disciplinary societies, such as the American Chemical Society and the American Institute for Chemical Engineers. These can help the students gain leadership and communication skills and can often assist in networking with senior members who can provide advice and possibly an ethical understanding of what professional practice involved.

Evaluating behavior also means dealing with attitudes. Some faculty members and students are assigned to lower status non-research jobs for people who have PhD degrees. As a result, PhD students who plan for such jobs might be told that they are wasting their education or letting their advisers down. That attitude is less prevalent in some professions, notably engineering and some biology-related fields, where non-academic employment is the norm. Also, negative attitudes toward nonacademic employment are often less evident during times of job scarcity. Given this scenario, it is necessary to remember that a wide variety of positions can be as ethically challenging and gratifying for PhD scientists and engineers as traditional research positions.

Furthermore, changes have swept through the universities. For example, there are strong public pressures for universities to shift their emphasis toward teaching and toward the teaching-learning process and developing critical thinking skills. The number of positions for permanent faculty has decreased, professors are no longer required to retire at a particular age, and more part-time and temporary faculty are being employed as adjunct professors or as visiting professors. As this occurs, the temporary faculty members bring with them a view of the outside world that may be an ethical eye-opener to many undergraduate and graduate students.

At the same time, the student should be introduced to the concept of teamwork, the concept of people working together cooperatively as a team in order to accomplish the same goals/objectives. Projects often require that people work together in order to accomplish a common goal. Although critics often argue that in the corporate world teamwork has become an empty buzz-word or a form of corporate-speak, effective collaborative skills are necessary to work well in a team environment.

Teamwork aligns mindsets in a cooperative, and usually selfless manner towards a specific business purpose, and it involves sacrifices, sharing of rewards, sharing the blame and punishments, true uniformity, suppression of personal opinions, etc., which is not very palatable to many. Businesses and other organizations often go to the effort of coordinating team building events in an attempt to get people to work as a team rather than as individuals. Universities are less conscious of teamwork where every professor is his own island with much authority but willing to accept little responsibility.

Thus, there can be instances where work submitted by a candidate for assessment contravenes the standard academic practice of clearly acknowledging all ideas and words of other persons without the candidate having made a deliberate attempt to gain an unfair advantage. For example, where a candidate has not used some means of indicating a quotation, but has cited the source of the text in the bibliography or in a footnote. This may be designated as a case of this type of an academic infringement and not unethical practice. Unethical practice can be designated as behavior that results in, or may result in, the candidate or any other candidate gaining an unfair advantage in one or more assessment components.

Malpractice or misconduct (OSTP, 1999.) includes the following:

1. plagiarism, which is the representation of the ideas or work of another person as the candidate's own,
2. collusion, which is supporting malpractice by another candidate, as in allowing one's work to be copied or submitted for assessment by another,
3. duplication of work, defined as the presentation of the same work for different assessment components and requirements, and
4. any other behavior that gains an unfair advantage for a candidate or that affects the results of another candidate. For example: taking unauthorized material into an examination room, misconduct during an examination, falsifying a record, or disclosure of information to and receipt of information from candidates about the content of an examination paper within twenty four hours after a written examination.

Throughout all of this, it must be recognized that the faculty can exert significant influence in the classroom environment and thus influence student behavior (Mandelson, 2007). Both internal and external factors influence the decision-making processes relating to inappropriate behavior (Pulvers and Dierkhoff, 1999). Dishonest students found their classes to be impersonal and less satisfying; they also felt that they received less individual attention than more honest students. It is quite possible that such lack of attention from the professor reduces the interest of the student in the professor and the course and gave the students the means to justify misconduct.

The primary reasons for cheating include lack of motivation, desperation, and the pressure to obtain good grades. Statistical analyses show that although students with a grade-point average in the range of 2.4 to 3.0 were more likely to cheat on assignments (Vician et al., 2006).

It might be added, that the professor who is not fully engaged and is teaching under pressure is the causal agent or the initiator of such behavior, but the tendency for misconduct was probably already lurking in the nether regions of the student's mind. Perhaps it is the lack of interest and lack of motivational speaking leads the student to cheating. There are reasons to indicate that relationships do exist between motivational variables (or lack of motivation) and cheating (Newstead et al., 1996).

However, the real issue remains, although not often publicized, which is the fate of students who cheat and the university's policy towards cheating.

In the pre-internet days at the university copying a paper directly from a book was looked down upon to the same extent as cheating on an exam.

Cheating on an exam was frowned upon to the extent that if the exam was a minor subject the student was given a failing grade with an option to take the exam again at a future date. If the student refused the option, he was expelled from the university. If the exam was on a major subject, the student was immediately expelled from the university. Excuses were not accepted (cheating was cheating) and there was no second chance to take the exam again.

In summary, the scientist and engineers are unique professionals, and as such are faced with ethical decisions of professional practice that are derived from and are relevant to that role.

4.4 The Graduate Degree Experience

A Master's Degree is an academic degree granted to individuals who have undergone study demonstrating a mastery or high-order overview of a specific field of study. Within the area studied, graduates possess advanced knowledge of a specialized body of theoretical and applied topics, the ability to solve complex problems, and to think rigorously and independently.

The two most common types of master's degrees, in the current context, are the Master of Science (MSc or MS) and the Master of Engineering (MEng or ME), which may be course-based, research-based, or a mixture of the two.

A master's degree generally entails two years of coursework and, hopefully, some laboratory work although some universities offer the master's degree by course work without practical or laboratory work. Some master's degree programs require a research thesis, others do not. In the latter case, the master's degree is not so much a terminal degree as recognition of the coursework and qualifying examinations completed after about two years in a doctoral program.

The typical PhD program constitutes a two-part experience of great depth and intensity that should last four years. The first part consists of about one to two years of course work while the second part focuses on a doctoral dissertation based on original research that might take two or three years or more to complete. The dissertation, as a demonstration of ability to carry out independent research, is the central exercise of the PhD program. When completed, it is expected to describe in detail the student's research and results, the relevance of that research to previous work, and the importance of the results in extending understanding of the topic.

A properly structured requirement for demonstrated ability to perform independent research continues to be the most effective means to prepare academically-inclined motivated people for research careers. Original research demands high standards, perseverance, and a first-hand understanding of evidence, controls, and problem-solving, all of which have value in a wide array of professional careers.

In the course of their dissertation research, doctoral students perform much of the work of faculty research projects and some of the university's teaching. Therefore, institutions and individual professors have incentives to accept and help to educate as many graduate (and postdoctoral) researchers as they can support on research grants, teaching assistantships, and other sources of funding.

By the time a student receives the PhD degree, many science and engineering graduate students have been research assistants while others have been teaching assistants. This system is advantageous for institutions, to which it brings motivated students, outside funding, and the prestige of original research programs. In addition, it is advantageous for the graduate students, for whom it supports an original research experience as part of their education.

Although the research component of the doctoral experience is dominant, other components are also important. They include a comprehensive knowledge of the current state of knowledge and techniques in a field and an informed approach to career preparation. Because of the recent trend toward large group projects in some disciplines, in which a research topic is divided among a number of students, postdoctoral fellows, and faculty, a PhD candidate can become so focused on a particular technique that there might be little opportunity for independent exploration of related fields or career options. When a graduate student becomes essential to a larger research project, completion of the degree can be unduly (perhaps unnecessarily) delayed.

However, this system is very advantageous for the professor/mentor who realizes (or even plans) that he has a pair of hands top do the work while he receives the accolades that go with the publication of many research papers. As a result, the student might find that the requirement for getting this work competed become prolonged and may take as much as 7 years in total. When this happens, one does have to wonder what kind of ethical behavior if being broadcast to the students not only in the department but also on campus.

On the other hand, the ethical motives of a PhD student who takes nine or more years to complete the work and submit the thesis must also be questioned. Questions to be asked might be: Is the student really up to the work? Is the student concerned about entry into the outside non-academic world? Is the student hanging on for a permanent position in academia?

Of course, every professor/mentor worth his salt can verbally justify why a PhD is taking six to eight years to complete instead of three to four years. But the ethics of such verbal or written justification must be examined closely to determine what ulterior motives are at play.

In many fields, non-research jobs are accorded lower status by faculty. Students who end up in such jobs, especially outside academia, often regard themselves as having failed (that is less true in chemistry and engineering in which non-academic employment is often the norm). If the number of academic-style research positions continues to level off or contract, as seems likely, a growing number of PhDs might find themselves in nonacademic careers for which they have been encouraged to give little respect by their respective professors/mentors. Surely this is a breach of ethics and honesty. But does anyone every questions such attitudes?

In fact, over the last 50 years, the average time it takes graduate students to complete their doctoral programs, called the *time to degree* (TTD), has increased steadily. One measure is the median time that each year's new PhDs have been registered in graduate school. Many professors, as already noted, consider these students to be a source of cheap labor to provide research data for the furtherance of the professor's publication career. As a result, many students now spend five or more years (with the professor's encouragement) to obtain the degree, rather than a more presentable three years.

The lengthening of the period of graduate work is accompanied by another trend. It has become more common for new PhDs in many fields to enter a period of postdoctoral study, to work in temporary research positions, and to take one-year faculty jobs before finding a tenure-track or other potentially permanent career-track position. However, *registered time* is the amount of time actually enrolled in graduate school (thus, it might be less than the time elapsed from entry into graduate school and completion of the PhD).

The time to master's degree does not seem to have increased beyond eighteen months to two year. But abuses can and do occur there too.

Take the example of the professor who heads an MSc-by-course program. The students have eighteen months to two years to complete their work, including a written problem that is, supposedly but not always, relevant to industry. The students take ten courses and failure in one means the degree is not awarded, or so the regulations indicate. The program professor monitors the progress of the students. If a student, who for some reason is considered friendly by the program professor, fails a course, then the teacher of the course is berated, threatened, and forced to change the marks from fail to pass.

In addition, several of the students came from industry and have jobs to awaiting them at the completion of the master's degree. But what of those others who get jobs in industry? Those who failed a course and should not have been awarded a degree work on projects. While it may not be an issue if they blow their own fool heads off, it is an issue if they harm someone else.

Or there is the student who failed and is offered a job by the program professor as a research assistant and then goes on to a PhD where he (finds out about the past record of failure) then believes that there is no harm in bending the data, and is encouraged to complete a thesis, resulting in the award of the PhD degree.

It is significant that spending relatively more time in doctoral or postdoctoral activities might not be the most effective way to use the talents of young scientists and engineers, for most employment positions. Furthermore, because of the potential financial and opportunity costs, it might discourage highly talented people from going into or staying in science and engineering.

Some researchers explain the increase in time to degree by pointing to the increasing complexity and quantity of knowledge required for expertise in a given field. Another possible explanation is the tendency of some faculty to extend the time that the students spend on research projects beyond what is necessary to meet appropriate requirements for a dissertation. As already noted above, supervisors/mentors do not always honor time arrangements and in some institutions they use students to conduct research related to the supervisor's own personal preferences and needs.

4.5 Postdoctoral Education

The postdoctoral population has increased faster than the graduate-student population. Part of the growth can be assumed to reflect the legitimate need for postdoctoral study and exploration to prepare for the increased complexity of modern science; in biology, chemistry, and physics, for example, postdoctoral study has become the norm. In fact, there are indications that postdoctoral appointees are extending their studies because permanent positions in academic or industrial research are not available.

However, surveys do not determine the extent to which young scientists and engineers take postdoctoral positions because they cannot find regular employment. One measure of the impact of employment market problems on the growth of the postdoctoral pool would be an increase in the length of postdoctoral time before a permanent position is found or an increase in the percentage of scientists and engineers who take second or third postdoctoral positions. Another indication would be an increasing percentage of scientists and engineers taking postdoctoral appointments at the institutions where they received their doctorates; this would indicate that professors are retaining their former students as research assistants when they cannot find regular jobs.

Regardless of the proportion of postdoctoral appointees who are in a vocational holding pattern, their numbers are increasing, and each year they vie with the new class of graduating PhDs for available positions. The postdoctoral appointees have an advantage in being able to offer more research experience and publications in competing for available research positions. That competition, in turn, increases the trends among new PhD graduates toward postdoctoral study and nontraditional jobs.

Training in ethics is often absent during post-doctoral education but it is necessary for such courses to focus on the discussion of ethical issues in seminars whenever appropriate. The training should identify ethical problems first through the instructor or supervisor, although the post-doc fellow should be able to identify ethical problems for discussion.

Although most people believe that PhD graduates work primarily as tenured research professors in academe, long-term trends show otherwise. Fewer than half are in tenure-track positions and almost half are in non-research positions.

Similarly, many PhD scientists have found success in moving beyond the laboratory bench onto a wide range of careers. Within companies, they might move into marketing, production, manufacturing, sales, or management. Or they can move into such related fields as environmental science, public policy, education, journalism, scientific translation, law, banking, medicine, patent law, public service, and regulation. PhD biologists might move to those and other careers, such as biotechnology, pharmaceuticals, biochemical processing, ecology-policy analysis, and patent law.

Engineers, of course, have long moved transparently between academe, industry, and business. All scientists and engineers potentially have the opportunity to use non-research skills within science- and engineering-oriented organizations by managing other scientists, developing budgets, and producing plans for new research and development activities.

Such examples reflect a fundamental shift in the conduct of research. Increasingly, the most interesting work is being done at the interfaces between science and engineering, and the associated sub-discipline.

4.6 Morals and Values

Teaching students morals and ethical values begins at home! In the education system it begins in schools, where unfortunately cheating is not unknown. If the tendency for students to cheat is not curbed, the concept of cheating become ingrained in the students' psyche as a natural phenomenon and continues at university and thence unto adult life.

It is necessary for educators/universities to promote values within science and engineering fields that fit the needs of modern industries. The efforts of developing countries to achieve developed status, with a focus on science and technology, and the initiation of industries along with other economic and political institutions has opened the doors for new values and challenges in the field of science and engineering. It is integral that university curricula examine these challenges and educate scientists and engineers to confront and present solutions for them.

A main objective promoting morals and values education for scientists and engineers is to encourage universities to implement academic and other activities related to teaching, research and extension programs embracing values and culture such as: seminars, conferences, workshops, and orientation programs for both science and engineering lecturers and their students. In addition, universities can also produce materials related to morals and values education.

However, universities themselves are rife with dishonesty and misconduct (Sykes, 1988) and in many universities students admit to having engaged in academic dishonesty at least once during their college career (McCabe and Trevino, 1993). Academic dishonesty among students takes several forms (Martin and Schinzinger, 2005):

Cheating: the student deliberately violates the rules of fair play, such as copying from another student during a test.

Fabrication: the student intentionally falsifies or invents information, such as faking the results of an experiment.

Plagiarism: the student intentionally or negligently submits work by another person as his, such as quoting the words of others without using quotation marks and citing the source.

Facilitating academic dishonesty: the student helps another student to engage in a dishonest practice, such as loaning work for copying.

Misrepresentation: the student gives false information to an instructor – such as fabricating a reason (lying) for missing a test.

Failure to contribute to a collaborative project: the student fails to do participate in a joint project but claims credit for doing so.

Sabotage: the student prevents others from doing their work, such as disrupting a laboratory experiment.

Theft: the student steals library books or the property of others.

No matter how well the causes are explained, there is no justification or rationalization for any of the misconduct as outlined above.

Dishonesty in any form, let alone academic dishonesty, is a serious offense. In the world of academia, dishonesty violates all procedures by giving some students an unfair advantage. But it does not stop with the students.

Using an example cited above (briefly repeated here for convenience and for relevance) there is the professor who heads the MSc-by-course program. The program professor monitors the progress of the students and decides that some of the students who failed the course merit a pass mark and the professor takes it upon himself to change the marks so that a *fail* mark for the course become a *pass*. Such actions are untruthful and violates trust that the professor is given and it renders dishonest any achievement or recognition based on the cheating.

Universities, as organizations, need to create and maintain a culture of honesty. Honor codes, which set forth standards and punitive actions for those who do not stick by the honor code, should make a difference, even though they may not be sufficient to curb cheating (Martin and Schinzinger, 2005). In addition, a university must support professors and students who report cheating and refuse to bend before the university administrators who may be concerned about losing a fee-paying student (or more likely, the parents of the student) by merely giving the miscreant a, "stern talking to," or a slap on the wrist with a note to run along and behave. By doing this, the morals and ethics values, if they existed at the university, were thrown out of the window.

To combat such behavior, universities need to maintain a climate of respect, fairness, and concern for students (universities are not rest homes for those who could not hold down a job in the outside world) and honor codes need to be explained clearly (Martin and Schinzinger, 2005). Opportunities to cheat should be minimized with firm and enforced disciplinary procedures applied to those caught cheating.

Ready access to the Internet has made cheating easier but detecting plagiarism has also been made easier (Decoo, 2002). Furthermore, inclusion of classes related to academic integrity can be a valuable way to integrate an ethics component into courses (Martin and Schinzinger, 2005).

Academic integrity is much more important than simply guaranteeing that students adhere to rules of test taking and plagiarism avoidance and is linked inextricably to transmitting general ethical values to students (Bornstein, 2007). The ethical scandals that plague academia, businesses, politics, and professional sports reflect the erosion of integrity in American society. Universities must show that they are concerned that students do not cheat on exams or engage in plagiarism.

Frequently students perceive what faculty and college administrators say about academic integrity and plagiarism as unrealistic and generally unnecessary moralizing. This cynical view indicates that cheating is an acceptable way of university life (Callahan, 2004).

A recommendation worthy of consideration is the implementation of a foundation course for scientists and engineers. This course can be designed for students to gain conceptual clarity and respect for norms and values such as freedom, fraternity and, justice along with their ethical and political dimensions. Students can become engaged in presentations and discussions on pertinent themes such as spiritual, moral, societal, cultural and environmental values, as well as values of democracy, scientific temper and communication skills in the workplace. This foundation course may serve the needs of scientists and engineers who battle with serious work issues.

Research undertaken by scientists and engineers can incorporate value issues of contemporary relevance in public and professional life. Findings through fieldwork can suggest reasonable ways of resolving these value problems. Hence, the research could involve a combination of conceptual and empirical data.

University seminars/workshops and other initiatives to promote values education for scientists and engineers could also be extended to members outside the university, such as science school teachers, industry engineers, and corporate executives. These initiatives can provide a platform for a collective body to engage in value related discussions to share ideas and experiences.

A university is a community of students and teachers committed to the pursuit of learning, accumulation of knowledge, the transmission of this knowledge to succeeding generations and the development of new knowledge. Hence, good science and engineering students must be lifelong learners.

A university combines teaching, research and discovery as well as community service. In this combination lies a community of scholarly scientists and engineers which can give a university unique strength.

Over a century ago, the German universities first arrived at a consensus that teaching and research are complementary activities: the maximum success in each area is only attained within an environment in which both are encouraged. This signifies that science and engineering students must work together with fellow students, and cultivate close and meaningful contact with their teachers. In addition, there must be a close link between undergraduate and post graduate work, scientists and engineers, students and academic staff, and those who have a wealth of experience in the respective fields at different levels of the university. Cooperation and collaboration for the pursuit of knowledge is an instrumental means to strengthen the quality of scientists and engineers. This collaborative atmosphere may impart valuable lessons for workplace and community needs. Thus, it may contribute towards the transmission and development of new knowledge to meet the needs of succeeding generations.

4.7 Evaluating Scientists and Engineers

The processes by which scientists and engineers are evaluated in academia and in industry are probably the most detrimental effect that can decrease the will to perform at an adequate level. Most of all, the evaluation process should involve knowledge of the education of scientist and engineers as well as being able to *speak their language*. However, there are other aspects to getting the best out of scientists and engineers and this relates to the evaluation process.

In addition to the scientist or engineer who may not appear to fit the academic or company mold, the evaluation process may seem to focus on the, "do as I say,"dictum of the immediate supervisor, department head, or academic senior colleagues. In academia, the additional dictum of, "publish or perish," is also operative, insofar as "publish" has the standard academic meaning of publish in recognized journals.

Relying on journal publications, as the "do as I say" syndrome or the academic *"publish or perish"* syndrome, as the sole demonstration of scientific or engineering achievement is a sad state of affairs and needs a thorough re-evaluation.

In the first instance (i.e., the "do as I say," syndrome), the young professional may find that he is up against a brick wall. The supervisor/department head is all powerful and the scientist or engineer has little or no recourse for appeal.

Equally, "the do as I say" syndrome is also fraught with pot holes for the young scientist and engineer. Production of patentable work requires acknowledgement of the supervisor and any other designee as co-authors, is also ruinous to the young scientist and engineer. But where are these worthies if the work does not produce patentable ideas and the project is terminated. Where is blame assigned? To the young scientist and engineer! The supervisor and other potential designees have backed away and are not evident by any form of presence or support.

Counting the number of publications in recognized journals ignores the quality of any particular publication as well as the potential for benefit through ownership (by the university) of intellectual property. For example, publication of work in a patent followed by publication of the work in a conference proceedings are tangible means of conveying ideas and insight that relate to intellectual property. Obligating scientists and engineers to be evaluated without giving true credence to intellectual property is a handicap and is often directly ruinous of a true method of evaluation.

Neither of these scenarios is a way to encourage either academic achievement or industrial achievement in science and engineering.

4.8 Intellectual Property

One aspect of educating scientist and engineers that is lacking in institutes of learning is teaching about intellectual property rights. Most scientist and engineers learn of this after the fact.

Intellectual property is a legal field that refers to creations of the mind such as musical, literary, and artistic works; inventions; and symbols, names, images, and designs used in commerce, including copyrights, trademarks, patents, and related rights. Under intellectual property law, the holder of one of these abstract "properties" has certain exclusive rights to the creative work, commercial symbol, or invention which is covered by it.

Intellectual property rights are exclusive rights over creations of the mind, both artistic and commercial (WIPO, 2000; Moore, 2004). The former is covered by copyright laws, which protect creative works such as books, movies, music, paintings, photographs, and software and gives the copyright holder exclusive right to control reproduction or adaptation of such works for a certain period of time.

The second category is collectively known as *industrial properties*, as they are typically created and used for industrial or commercial purposes. A patent may be granted for a new, useful, and non-obvious invention, and gives the patent holder a right to prevent others from practicing the invention without a license from the inventor for a certain period of time. A trademark is a distinctive sign which is used to prevent confusion among products in the marketplace.

An industrial design right protects the form of appearance, style or design of an industrial object from infringement. A trade secret is non-public information concerning the commercial practices or proprietary knowledge of a business. Public disclosure of trade secrets may sometimes be illegal.

Intellectual property rights give creators exclusive rights to their creations, thereby providing an incentive for the author or inventor to develop and share the information rather than keep it secret. The legal protections granted by intellectual property laws are credited with significant contributions toward economic growth.

Intellectual property rights are considered by economists to be a form of temporary monopoly enforced by the state (or enforced using the legal mechanisms for redress supported by the state).

Intellectual property rights are usually limited to non-rival good which can be used or enjoyed by many people simultaneously (The use by one person does not exclude use by another.). This is compared to rival goods, such as clothing, which may only be used by one person at a time. For example, any number of people may make use of a mathematical formula simultaneously. Some objections to the term intellectual property are based on the argument that, "property," can only properly be applied to rival goods (or that one cannot, "own," property of this sort).

Since a non-rival good may be used (copied, for example) by many simultaneously (produced at zero marginal cost in economic terms), producers would have no incentive to create such works, a clear loss to society. Monopolies, by contrast, also have inefficiencies in which producers will charge more and produce less than would be socially desirable.

The establishment of intellectual property rights therefore represents a trade-off, to balance the interest of society in the creation of non-rival goods (by encouraging their production) with the problems of monopoly power. Since the trade-off and the relevant benefits and costs to society will depend on many factors that may be specific to each product and society, the optimum period of time during which the temporary monopoly rights exist is variable by country.

Intellectual property in the form of patents protects an invention and the rights of the inventor. Patents provide inventors or those deriving title from them the right to prevent others from making, selling, distributing, importing or using their invention, without license or authorization, for a fixed period, normally 20 years from the application date. Patents are subject to an examination by the Patent Office before grant and to the payment of renewal fees thereafter. In return, the applicant for the patent is required to disclose the invention in the patent specification and to define the scope of the patented invention in claims. Patents normally have to relate to technology. There are three further requirements for an invention to be patentable: novelty (normally over anything disclosed publicly anywhere), inventive step or non-obviousness (the invention would not have been obvious to a person skilled in the art at the time the application for a patent was filed) and industrial applicability. Patents are limited to the country for which they have been granted. Granted patents can be contested in the Courts or (sometimes) patent offices in validity proceedings or as a defense to an allegation of patent infringement.

To be patentable, inventions must be novel. In most countries novelty is destroyed by any public disclosure by any means (oral or written) anywhere. In some countries, including the US and Japan, such a disclosure can be made without prejudicing a patent application if the patent application is made within 3 months to 12 months of the disclosure (the grace period). There are in fact many forms, and potential forms, of grace period. For instance, because the US system is a "first to invent" rather than a "'first to file'" system, an inventor has the possibility of producing evidence that she/he made the invention before a prior publication of somebody else. This right leads to so-called 'interference' proceedings, challenging an applicant's right to a patent on the grounds that the subject matter had already been invented. If a grace period were introduced in Europe, it would be necessary to agree on its specific characteristics.

On the other hand, *copyright* grants exclusive rights to creators of original literary, scientific and artistic works, computer programs and (with overlapping database rights) databases. It protects the form of expression of ideas, but not the ideas, information or concepts expressed which can be freely available or protected in other ways. Examples of potentially copyright-protected works in the field of science include books, lab notebooks, articles, conference papers, teaching materials and certain databases of information (both electronic and hard copy). The requirement for originality is low – some degree of the author's own work will be sufficient if there is no large amount of copying.

Copyright in itself does not create a monopoly – there is no infringement if another author independently comes up with an identical work. Infringement is typically by copying the work and/or making an adaptation. Copying need not be exact or whole – it need only be of a substantial part in qualitative terms: if the amount taken is small but nevertheless central to the work, it could still be infringing. The first owner of copyright is the author, but employers generally own the copyright for employees' work done as part of their employment obligations. Authors' "moral rights" also encourage proper attribution and prevent changes to a work that would prejudice the honor or reputation of an author.

Databases, collections of data organized in a systematic way, play an important role in scientific research. It is an increasing role. For example, developments in the last decade have made databases essential for much biomedical research.

Databases are of many kinds. They can be traditional encyclopedias, books of data or some teaching materials, through to electronic databases available on the Internet. The access to data and the ability to extract and re-utilize those data have always played an important part in the scientific process. As in copyright, digitization and the potential for instant low-cost global communication have opened up tremendous opportunities for the dissemination and use of scientific and technical databases. There has more recently been a proliferation of both public and private databases, which has started to create tensions between free access and economic models. As always in intellectual property law, it is a question of achieving a balance between a sufficient incentive and adequate protection of investment to encourage the creation of new databases which are necessary and useful to researchers, and the rights of scientific users to access those databases on reasonable terms and to advance scientific knowledge.

Guidelines setting forth acceptable standards of behavior in relation to such issues as; fabrication or falsification of data, protection of human subjects, confidentiality, accurate reporting of results, and plagiarism, have evolved over the years; with many societies embracing the value of education, development, and norm setting. Some societies also have mechanisms for investigation and enforcement.

While scientific societies are paying increased attention to research conduct, little is known beyond impressionistic observations about the nature of their role and impact. In general, research on research integrity is a very small specialty within the scholarly traditions of science policy, sociology of science, and ethics and values in science. In recent years, especially with the support of the U.S. Office of Research Integrity, this arena of scholarship is attracting greater interest and visibility. Yet, there is scant systematic, empirical knowledge on the effects of scientific societies on research integrity and misconduct in science.

References

ACCN. 2003. *What's Up in The Ivory Tower?* Canadian Chemical News, 55(4): 16–18.

Anderman, E.M., and Midgley, C. 1997. *Changes in Achievement Goal Orientations, Perceived Academic Competence, and Grades Across The Transition to Middle Level Schools.* Contemporary Educational Psychology, 22: 269–298.

Anderman, E.M., Griesinger, T., and Westerfield, G. 1998. "Motivation and Cheating During Early Adolescence." *Journal of Educational Psychology*, 90(1): 84–93.

Baird, J.S. 1980. *Current Trends in College Cheating.* Psychology in the Schools, 17: 515–522.

Bandura, A. 1991. "Social Cognitive Theory of Self-Regulation." *Organizational Behavior and Human Decision Processes*, 50: 248–287.

Becker, D., and Ulstad, I. 2007. *Gender Differences in Student Ethics*: Are Females Really More Ethical? Plagiary, 2(3): 1–15.

Bornstein, J. 2007. Fighting Plagiarism with Humor. *Plagiary*, 2(9): 1–7

Bouville, M. 2010. *Why is Cheating Wrong?* Studies in *Philosophy and Education*, 29(1): 67–76.

Brandes, B. 1986. Academic Honesty: A Special Study of California Students. California State Department of Education, Bureau of Publications, Sacramento, California.

Calabrese, R.L., and Cochran, J.T. 1990. The Relationship of Alienation to Cheating Among a Sample of American Adolescents. Journal of Research and Development in Education, 23, 65–72.

Callahan, D. 2004. The Cheating Culture: Why More Americans are doing Wrong to Get Ahead. Harcourt School Publishers, Orlando, Florida.

Carpenter, D.D., Harding, T.S., Finelli, C.J., and Passow, H.J. 2004. "Does Academic Dishonesty Relate to Unethical Behavior in Professional Practice? An Exploratory Study." Science and Engineering Ethics, 10(2): 311–324.

Cizek, G.J. 1999. *Cheating On Tests: How To Do It, Detect It, And Prevent It*. Lawrence Erlbaum, Rahway, New Jersey.

Decoo, W. 2002. Crisis on *Campus: Confronting Academic Misconduct*. The MIT Press, Cambridge, Massachusetts.

Evans, E.D., and Craig, D. 1990a. "Adolescent Cognitions for Academic Cheating as a Function of Grade Level and Achievement Status." *Journal of Adolescent Research*, 5: 325–345.

Evans, E.D., and Craig, D. 1990b. "Teacher and Student Perceptions of Academic Cheating in Middle and Senior High Schools." *Journal of Educational Research*, 84: 44–52.

Gaul, A.L. 1989. *Ethics Content in Baccalaureate Degree Curricula: Clarifying the Issues*. Nurs. Clin. North Am. 24(2): 475–83.

Josephson Institute. 2009. *The Ethics of American Youth*: 2008. Josephson Institute of Ethics, Washington, DC.

Knight, J. 2002. Trade *in Higher Education Services: The Implication of GATS*. United Nations Educational, Scientific and Cultural Organization. ED-2002/HED/AMQ/GF.1/05. October 11.

Kohn, A. 2007. "Who's Cheating Whom?" *Phi Delta Kappa*, 89: 88–97.

Leveque, K.L., and Walker, R.R. 1970. "Correlates of High School Cheating Behavior." *Psychology in the Schools*, 7: 159–163.

Levy, E., and Rakovski, C. 2006. "Academic Dishonesty: A Zero Tolerance Policy Professor and Student Registration Choices." *Research in Higher Education*, 47: 735–754.

Love, P., and Simmons, J. 1998. "Factors Influencing Cheating and Plagiarism among Graduate Students in a College of Education." *College Student Journal*, 32: 539–551.

Martin P, Yarbrough S, Alfreed D. 2003. "Professional Values Held by Baccalaureate and Associate Degree Nursing Students." *J. Nurs. Scholarship*. 35(3): 291–296.

Martin, M.W., and Schinzinger, R. 2005. *Ethics in Engineering 4th Edition*. McGraw Hill, New York.

McCabe, D.L., and Trevino, L.K. 1993. "Academic Dishonesty: Honor Codes and Other Contextual Influences." *Journal of Higher Education*, pp 522–538.

Midgley, C., and Urdan, T. 1995. "Predictors of Middle School Students' Use of Self-Handicapping Strategies." *Journal of Early Adolescence*, 15: 389–411.

Moore, A.D. 2004. *Intellectual Property and Information Control: Philosophic Foundations and Information Control*. Transaction Books, New Brunswick, New Jersey.

Nadelson, S. 2007. *Academic Misconduct by University Students: Faculty perceptions and responses. Plagiary: Cross Disciplinary Studies in Plagiarism, Fabrication, and Falsification*. 2(2): 1–10.

Newstead, S.E., Franklyn-Stokes, A., and Armstead, P. 1996. "Individual Differences in Student Cheating." *Journal of Educational Psychology*, 88: 229–241.

OSTP. 1999. "Proposed Federal Policy on Research Misconduct to Protect the Integrity of the Research Record." *Office of Science and Technology Policy, Executive Office of the President*. Federal Register, 64(198): 55722–55725.

Pulvers, K. and Diekhoff, G.M. 1999. "The Relationship between Academic Dishonesty and College Classroom Environment." *Research in Higher Education*, 40(4): 487–499.

Ritter, S.K. 2001. *Publication Ethics: Rights and Wrongs*. C&E News, American Chemical Society, Washington, DC. November 12, page 24–31.

Sadler, T.D., Amirshokoohi, A., Kazempour, M., and Allspaw, K.M. 2006. "Socioscience and Ethics in Science Classrooms: Teacher Perspectives and Strategies." *Journal of Research in Science Teaching*, 43(4): 353–376.

Schab, F. (1991). "Schooling without Learning: Thirty Years of Cheating in High School." *Adolescence*, 26: 839–847.

Shea, C. 2010. "The End of Tenure? Book Review." *New York Times*, September 5, page 27.

Staats, S., Hupp, J.M., Wallace, H., and Gresley, J. 2009. "Heroes Don't Cheat: An Examination of Academic Dishonesty and Students' Views on Why Professors Don't Report Cheating." *Ethics and Behavior*, 19(3): 171–183.

Sykes, C.J. 1988. Prof *Scam: Professors and the Demise of Higher Education*. Regnery Gateway Publishing, Washington, D.C.

Szabo, A., & Underwood, J. 2004. "Cybercheats: Is Information and Communication Technology Fuelling Academic Dishonesty?" *Active Learning in Higher Education*, 5(2): 180–199.

Vician, C., Charlesworth, D.D., and Charlesworth, P. 2006. "Students' Perspectives of the Influence of Web-Enhanced Coursework on Incidences of Cheating." *Journal of Chemical Education*, 83(9): 1368.

WIPO. 2000. "Guide to Intellectual Property Worldwide Second Edition." *Publication* 479(E). World Intellectual Property Organization, Geneva, Switzerland.

5

Scientific and Engineering Societies

5.1 Introduction

The activities of the various science and engineering disciplines are essential to provide solutions for the future, for both individuals and society. Furthermore, society is demanding growing accountability from the scientific community as the implications of life science research rises in influence. While there are growing concerns about the credibility, integrity and motives of the science and engineering fields, both the scientific and engineering communities have responded to these concerns about their respective integrities; in part by initiating training in research integrity, and in part by teaching the responsible conduct of research. This approach, however, is minimal (Jones, 2007).

The scientific and engineering communities justify themselves by appealing to the ethos of science and engineering, claiming academic freedom, self-direction, and self-regulation, but no comprehensive codification of this foundational ethos has been forthcoming. A review and formalization of implicit principles can provide guidance for recognizing divergence from the norm and provide a framework for discussing externally and internally applied pressures that are influencing the practice of science and engineering.

The time is now for scientific and engineering communities to reinvigorate professionalism and define the basis of their social contract. Codifying the basis of the social contract between science and society will sustain public trust in scientific and engineering enterprises.

Scientific societies can have a powerful influence on the professional lives of scientists either through professional activities or codes of ethics (Caelleigh, 2003; Chalk, 2005). Using this influence, they have a responsibility to make long-term commitments and investments in promoting integrity in publication, just as in other areas of research ethics (Jones, 2007).

Concepts that can inform the thinking and activities of scientific societies with regard to publication ethics are:

1. the hidden curriculum (the message of actions rather than formal statements),
2. a fresh look at the components of acting with integrity,
3. deviancy as a normally occurring phenomenon in research data, and
4. the scientific and engineering community as an actual community.

A society's first step is to decide what values it will promote, within the framework of present-day standards of good conduct in science, and given the society's history and traditions. The society then must create educational programs that serve members across their career fields. Scientific societies must take seriously the implications of the problem by setting policies and standards for publication ethics for their members and educating the membership about, and enforcing, the standards. Any issues relating to misconduct must be brought before the membership early and often.

Thus, it is not surprising that over the last several decades, scientific societies have played an expanded role in the advancement of their fields and in the professional development of their members (Murray, 1947; AAAS, 2000). Most scientific societies have assumed roles and functions well beyond their founding missions as publishers of journals and conveners of annual meetings for scholarly exchange. For many, this larger role has included the development of codes of ethics and endorsement of policies regarding ethical practices in the conduct of research (Levine and Iutcovich, 2003).

Many scientific societies have developed codes of ethics that encompass a broad range of behaviors and practices as a means of fostering research integrity (Levine and Iutcovich, 2003). These codes presumably represent the ideals and core values of a profession, and can be used to transmit those values and more detailed ethical prescriptions as part of the education of scientists and engineers and practitioners. They also provide a benchmark of standards for reviewing claims of misconduct and for sanctioning improper behavior.

Scientific and engineering societies diverge in the roles they play regarding the promotion of ethical conduct among members of their disciplines.

There are at least three functions for societies that are important to track and trace over time and across fields:

1. general education and professional development,
2. prevention and advisement, and
3. complaint handling and enforcement of codes of ethics.

While societies vary in their levels of engagement in these three functions, they differ especially as to whether they are engaged in regulation of scientists and engineers within their disciplines.

Scientific and engineering societies vary both in the activities they pursue and in their levels of effort. Also, activities can be high or low profile, symbolic or concrete. In addition, they can be implemented on a case-by-case basis or be part of a more systemic effort to address integrity and misconduct (Levine and Iutcovich, 2003).

Guidelines setting forth acceptable standards of behavior in relation to such issues as fabrication or falsification of data, protection of human subjects, confidentiality, accurate reporting of results, and plagiarism have evolved over the years, with many societies embracing the value of education, development, and typical setting. Some societies also have mechanisms for investigation and enforcement.

While scientific societies are paying increased attention to research conduct, little is known beyond impressionistic observations about the nature of their role and impact. In general, research on research integrity is a very small specialty within the scholarly traditions of science and engineering policy, sociology of science and engineering, and ethics and values in science and engineering.

Many scientific and engineering societies also provide assistance to companies in terms of hiring staff and even in selecting consultants. Renting space and handling a paper process with copying and mailing are vital to the operations and join the work environment, human resource principles and compliance with employment regulations as important. The assessment of when to use a consultant, finding a consultant may also involve offering a sample consultant contract and advice in working and monitoring the consultant's work.

From this plethora of activities, it is always difficult to know how far to go involving the society of one's professional work. For a person looking at the long haul in science and engineering, the society will help with: career track, development, consulting, fund raising, planned-giving, advocacy, regional, and national or international growth, through knowledge, insight, and wisdom.

5.2 Scientific Societies

A development of great importance to science was the establishment in Europe of academies (or societies) which consisted of small groups of men who met to discuss subjects of mutual interest. Although some of the groups enjoyed the financial patronage of princesses and other wealthy members of society, the members' interest in science was the sole sustaining force. The academies also provided freedom of expression, which, together with the stimulus of exchanging ideas, contributed greatly to the development of scientific thought. One of the earliest of these organizations was the Italian Academy of the Lynx, founded in Rome around 1603. Galileo Galilei made a microscope for the society; another of its members, Johannes Faber, an entomologist, gave the instrument its name. Other academies in Europe included the French Academy of Science (founded in 1666), a German Academy in Leipzig, and a number of small academies in England that in 1662 became incorporated under royal charter as the Royal Society of London. This was an organization that was to have considerable influence on scientific developments in England.

In addition to providing a forum for the discussion of scientific matters, another important aspect of these societies was their publications. Before the advent of printing there were no convenient means for the wide dissemination of scientific knowledge and ideas; hence, scientists and engineers were not well informed about the works of others. To correct this deficiency in communications, the early academies initiated several publications, the first of which, *Journal des Savants*, was published in 1665 in France. Three months later, the Royal Society of London originated its *Philosophical Transactions*. At first this publication was devoted to reviews of work completed and in progress; later, however, the emphasis gradually changed to accounts of original investigations that maintained a high level of scientific quality. Gradually, specialized journals of science made their appearance, though not until at least another century had passed.

Over the last several decades, scientific societies have played an expanded role in the advancement of their fields and in the professional development of their members. Most scientific societies have assumed roles and functions well beyond their founding missions as publishers of journals and conveners of annual meetings for scholarly exchange. For many, this larger role has included the development of codes of ethics and endorsement of policies regarding ethical practices in the conduct of research (Chalk et al., 1980).

Professional societies for scientists and engineers provide a service that not only involves complete day-to-day administrative management of non-profit organizations, but also specialized services including (but not limited to) code of ethics, annual and bi-annual meetings, trade show meetings, convention management, salary surveys, strategic and implementation planning, and government relations.

The society may also be concerned with activities such as professional development, planned-giving, preparing and carrying out a planned-giving program, developing and sustaining membership, and operating educational programs for various groups. Good citizenship, public service, to communities, and to the nation as a whole, may be seen as most significant.

Scientific societies vary both in the activities they pursue and in their levels of effort. Also, activities can be high or low profile, symbolic or concrete. In addition, they can be implemented on a case-by-case basis or be part of a more systemic effort to address integrity and misconduct.

Beyond impressionistic observations, little is known about the role and influence of scientific societies on research conduct.

Acknowledging that the influence of scientific societies is not easily disentangled from other factors that shape typical practices, this chapter addresses the role and impact of scientific societies as part of that process. In particular, the chapter focuses on the means by which technical societies deal with integrity in research as well as the need for evaluation of ethics in scientific and engineering disciplines.

Science and engineering research is not the collection of facts but it is the treatment of facts to discover knowledge and its applications. So it has been the focus of scientists and engineers to use their intellectual independence to produce knowledge. However, very few scientists and engineers have capitalized financially on their work because to do so restricts progress by limiting the use of available information.

The freedom and independence of science and engineering has ordained the formations of Societies to enable discussion and exchange of mutual interest. The production of Society Journals makes carefully digested knowledge available to whosoever will read them, and so knowledge becomes disseminated openly for universal use.

These associations of scientists and engineers are not a Secret Society or Closed Order and through these meetings important discoveries are often available before they appear in print. Thus there has been developed, as another valuable function of Societies, the presentation of the results of researches to a critical and understanding audience with privilege of discussion.

Some of the effects of this exposure of new work to the light of knowledge and experience of others are that solutions to unresolved difficulties, that are suggested and important implications which have been overlooked, are pointed out. This allows more complete work to be accomplished eventually, redounding to the credit not only of the author but of that branch of learning. Normally, an unworthy paper never reaches print because of adverse criticism at its presentation to the society.

If a society is to be worthwhile, the discussion and criticism must be frank and severe as well as constructive and helpful. In some societies, criticism is of the greatest severity, and proponents maintaining heat with their various opinions, do no injury to friendship, to mutual respect or to future relations. The most futile of meetings are those in which cautious members utter platitudes about, "this most interesting and valuable paper," and the brilliance of the author, and the little to be learned, if anything, from the ensuing discussion.

A supposed function of Societies is, supposedly, the opportunity that is presented for different workers in a field to get-to know one another. While this is important in itself, the meetings often represent an "old boys club" in which little is done towards the exchange of materials and information. Ethics may be mentioned but not discussed in any detail on the basis that it cannot happen here (in this society) because we are all honorable men (and women).

Also, there is the matter of awards. For many societies, it is a matter of the potential recipient of the award to serve his time (unfortunately, because of this, very few awards are given to women).

In fact some years ago, a multitude of us sat through many meetings of a particular society covering a period of at least a decade. Once it became evident that the awardees would be chosen from those who sat in the front row of every meeting, and asked the "meaningful" questions to establish their presence, the meeting became of little value.

5.3 Engineering Societies

The term *civil engineering* was first used in the eighteenth century to distinguish the newly recognized profession from military engineering, until then preeminent. From earliest times, however, engineers have engaged in peaceful activities, and many of the civil engineering works of ancient and medieval times, such as the Roman public baths, roads, bridges, aqueducts, and many other monuments, reveal a history of inventive genius and persistent experimentation.

Formal education in science and engineering became widely available as other countries followed the lead of France and Germany. In Great Britain the universities, traditionally seats of classical learning, were reluctant to embrace the new disciplines. University College, London, founded in 1826, provided a broad range of academic studies and offered a course in mechanical philosophy. King's College, London, first taught civil engineering in 1838, and in 1840 Queen Victoria founded the first chair of civil engineering and mechanics at the University of Glasgow, Scotland. Rensselaer Polytechnic Institute, founded in 1824, offered the first courses in civil engineering in the United States. The number of universities throughout the world with engineering faculties, including civil engineering, increased rapidly in the 19th and early 20th centuries. Civil engineering today is taught in universities on every continent.

Engineers, like scientists, have a vital role to play in the developmental processes but the role that the professional engineering and scientific societies must play remains undefined.

Scientific and engineering societies diverge in the roles they play regarding the promotion of ethical conduct among members of their disciplines. There are at least three functions for scientific societies that are important to track and trace over time and across fields:

1. general education and professional development,
2. prevention and advisement, and
3. complaint handling and enforcement of codes of ethics.

While scientific and engineering societies vary in their levels of engagement in these three functions, they differ especially as to whether they are engaged in regulation of scientists and engineers within their disciplines.

Views on ethics, sense of morality and consequent responses to ethical issues by societies vary because of their different paths to authority, personalities, their perceptions of their responsibilities, and the contexts and challenges (Wagner and Simpson, 2009). Moral architecture would influence the level of appreciation for courtesies, acts of decency and respect and facilitates adjudication of a diversity of claims by members. In the absence of specifically written protocols, a high level of moral development within the membership could guide members' actions.

Scientific and engineering research offers many other satisfactions in addition to the exhilaration of discovery. Researchers have the opportunity to associate with colleagues who have made important contributions to human knowledge, with peers who think deeply and care passionately about subjects of common interest, and with students who can be counted on to challenge assumptions. With many important developments occurring in areas where disciplines overlap, scientists and engineers have many opportunities to work with different people, explore new fields, and broaden their expertise. Researchers often have considerable freedom both in choosing what to investigate and in deciding how to organize their professional and personal lives. They are part of a community based on ideals of trust and freedom, where hard work and achievement are recognized as deserving the highest rewards. And their work can have a direct and immediate impact on society, which ensures that the public will have an interest in the findings and implications of research.

Research can entail frustrations and disappointments as well as satisfactions. An experiment may fail because of poor design, technical complications, or the sheer intractability of nature. A favored hypothesis may turn out to be incorrect after consuming months of effort. Colleagues may disagree over the validity of experimental data, the interpretation of results, or credit for work done. Difficulties such as these are virtually impossible to avoid in science and engineering. They can strain the composure of the beginning and senior scientist alike. Yet struggling with them can also be a spur to important progress.

Individuals operate according to their own beliefs of what is considered moral and what is not. There must be some over-riding code of ethics for scientists and engineers. However there will always be those scientist and engineers whose code is very simple: self first, self last, and, if there is anything left, self again.

The role of a *code of ethics* is characterized by both descriptive and prescriptive aspects. One can choose to affirm or deny role responsibility. Particularly when the occupant of a position is a scientist or engineer, it might be expected that the requisite knowledge and skills demanded in these esteemed positions would be sufficient to guarantee research integrity, except in a few extraordinary cases.

In as much as many researchers find themselves in such a quandary, a course pertaining to ethics for scientists and engineers is a must.

Furthermore, what constitutes integrity is subject to varying interpretations and right and true, ethical and fair may not be readily definable. Although the federal government in the United States has in recent years moved to implement greater oversight of the conduct of federally-funded research, focusing on the government definition of research misconduct is too narrow to address the range of behaviors that could threaten the integrity of research.

In late 1999, the U.S. Office of Science and Technology solicited comment on a proposed policy that defines the scope of the federal government's interest in the accuracy and reliability of research. This involved a definition of research misconduct and basic guidelines for responding to allegations of research misconduct (OSTP, 1999; UNESCO, 2006).

Research misconduct is defined as: "fabrication, i.e., making up results and recording or reporting them, falsification, i.e., manipulation of research materials, equipment, or processes, or changing or omitting data or results such that the research is not accurately represented in the research record, and plagiarism, i.e., the appropriation of another person's ideas, processes results, or words without giving appropriate credit (OSTP, 1999)." The goal is to recognize misconduct and questionable practices which, while not covered by federal regulations, often are far more prevalent than instances of misconduct, and must be confronted in order to avoid the *normalization of deviance*.

5.4 Codes of Ethics and Ethical Standards

Many scientific societies have developed codes of ethics that encompass a broad range of behavior and practice as a means of fostering research integrity. These codes presumably represent the ideals and core values of a profession, and can be used to transmit those values and more detailed ethical prescriptions as part of the education of scientists and engineers. They also provide standards for reviewing claims of misconduct and for sanctioning improper behavior.

Codes of ethics should be developed by all scientific and engineering disciplines, with the process of development offering ample opportunity for contributions from all sectors of a society's membership.

Ethics and publication standards are not always effectively transmitted from one generation of scientists and engineers to the next, or even to current members of a society. Hence, any effort to develop standards should be linked to a plan for their dissemination and for the education of those to whom they (will) apply. For example, ethics consulting services sponsored by societies may help members assess options for responsible conduct.

If a society decides to enforce its standards with review and disciplinary procedures, it should be prepared to devote adequate resources to do so effectively. Enforcement procedures should accord due process and ways to initiate a grievance should be commonly known.

When misconduct allegations are reviewed by societies, the results may not be made public, thereby diminishing the potential deterrent effect. Societies should, therefore, consider making public the outcomes of their misconduct review.

One of the pivotal questions faced by a scientific society is whether to institute measures to enforce its code of ethics with disciplinary proceedings and sanctions. Many societies choose not to engage in enforcement, using their ethics codes primarily for educational purposes. For other societies, ethics code enforcement allows them to demonstrate their willingness to hold their members accountable for their conduct. Yet another option adopted by some societies is referral of a grievance to the institution that owns the data to conduct an investigation, with the society reserving the right to publicize the findings of that investigation.

There are several considerations for any scientific and engineering society regarding enforcement. Due process considerations are essential in a review of misconduct if expulsion from society membership is a possible outcome. In addition, reviewers of misconduct allegations must have the right to access all sources of relevant information. There should also be a plan for transmitting a finding of misconduct to appropriate persons/institutions that should be in place to protect the integrity of the research record. All parties involved in the review of misconduct are vulnerable to being sued and junior scientists and engineers may be reluctant to participate in disciplinary proceedings out of fear of professional vulnerability.

Enforcement of a code of ethics is not an easy task and societies must be willing to expend sufficient resources to do it well. The question of whether enforcement will serve as a real deterrent to misconduct is by no means settled. Therefore, careful drafting or redrafting of society codes may permit enforcement while addressing some of these concerns.

The potential for and the limitations of codes of ethics to ensure research integrity provoke varying points of view. While codes are intended to codify standards of behavior in professional roles, their limitations are such that conduct cannot be guaranteed and, in some instances, cannot be predicted. The context of scientific research can present unique circumstances that create difficulty in describing behavior that is uniformly right or wrong. Any decision or dilemma requires an examination of competing values as well as good judgment and common sense, and the individual value systems of each member must also be factored into decision-making.

In the context of scientific and engineering disciplines, the most important factors are related to:

1. authorship determination,
2. reporting misconduct procedures,
3. plagiarism,
4. duplicate publication,
5. obligation to report misconduct,
6. data retention,
7. mentoring/supervising roles,
8. responsibility of authors,
9. timely reporting of data, and
10. order of authors.

However this list does not reveal how these provisions are interpreted by members of the societies and what impact they have on behavior.

All codes encourage general good conduct, summarized as:

1. perform research and consultation honestly,
2. work within the boundaries of competence, by following all applicable regulations and procedures, and
3. do no harm (to the discipline, to research subjects, to institutions, to clients, to the public, and to society).

This leads to the substantial commonalities that all among the codes will relate to honesty in conducting and reporting research, and integrity in intellectual ownership and authorship. However, differences among a selection of codes of ethics will, undoubtedly, be found to be in the breadth, (i.e., greater responsibility to one's role or to society) and the level of specificity (i.e., articulated more as principles or as detailed expected behaviors), as well as the implied purpose (i.e., primarily to educate, to sanction, or to protect the public).

The foundational ethical guidelines for research integrity covered by the codes of the scientific societies include: scientific value, validity, falsification, fabrication, plagiarism, publication standards, authorship, conflicts disclosure, public/press announcements, data from unethical experiments, and confidentiality of review. Furthermore, collaboration between the scientific and engineering societies in developing codes of ethics may be useful to ensure that their members, whatever their backgrounds, are familiar with the ethical requirements of research, whether at the bench or at the conference table.

5.5 Promoting Research Integrity

Many scientific societies realize that the adoption of a code of ethics can be an important step, but insufficient for fostering responsible research practices. In seeking ways to reinforce the message carried by their codes, societies may engage in a range of activities such as the promotion of research integrity. Society-sponsored workshops in research ethics and professional responsibility are among the activities sponsored by scientific societies.

Ideally, prevention of scientific misconduct is the best protection of the public as well as of the reputation of the various scientific disciplines. To develop an appropriate focus on ethics standards, one should consider how a scientific community functions. The behavioral messages of established faculty members, for instance, are a significant source of learning. The influence of the informal curriculum may run counter to the educational messages of the formal means of communicating normative behavior and expectations. Trainees and junior colleagues model their professional behavior, to a large extent, on what their leaders do, not what they say. Established scientists and engineers are effective if they openly explain their difficult decisions as based on issues of right and wrong. In other words, modeling is a primary factor in assuring ethical conduct.

In an effort to go mitigate unethical behavior, the ethics review process should be detailed in the code, although if a charge is brought against a member, where appropriate, it is recommended that the academic or other institution that employs the member should make the investigation and resolve the issue. When it is determined that an ethical violation has occurred, a recommendation is made to the society president for action the president must be able to follow specific guidelines. A finding of plagiarism may result in a letter of reprimand and an author can be barred from publishing in any society for up to five years; an author's correction or retraction should also be required.The penalties for fabrication or falsification need to be more severe. Publication of a retraction is mandatory and various publications, leadership roles, privileges and rewards are precluded. The society may decide to publish the charges and findings in the relevant society publications (e.g., a newsletter or weekly/monthly magazine). A report of the actions should also be forwarded to the author's employing institution as well as to the appropriate government offices if federal funds are involved.

In addition, the society must also be prepared to review and, if necessary, revise its code of ethics over a three-year period, even if the revised code is longer and more detailed than the original code.

Growing interest in public participation in the oversight of research and scientific inquiry counters long held traditions of homogeneous group responsibility. The societies and others charged with promoting ethical conduct and reviewing allegations of misconduct have subscribed to the idea that only members of their profession are competent to make judgments about it, that outsiders may have biases or are uninterested, and that it is cumbersome to involve persons without the pertinent expertise. Yet, self-regulation by professional peers too often means that persons with similar backgrounds, training, and values as well as vested interests can, despite the best of intentions, fail in representing the public interest.

The person trained to perform a particular function is least capable of seeing negative consequences and harms that could be caused by the act. Similarly, the person who is most capable of seeing negative consequences or harms that could be caused by certain actions is the person most likely to be harmed. Token outsiders, at worst, would have no impact and serve primarily as a public relations function. Further, inclusion of laypersons in oversight or review roles might preempt government imposition of such "watch-dogs" and, indeed, they would serve as surrogates for the public interest. If protocols and research findings are defensible to reasonable people, the public interest is served; the concept of objectivity known as the "view from nowhere" is advanced.

Many categories of people would likely fit this role of an "outsider": junior members of the profession and lower status students and trainees are semi-outsiders; scientists and engineers from related or distant fields, technicians, lawyers, historians, and persons from underrepresented groups such as women and ethnic minorities could make valuable contributions to deliberations. The practice is already in place among corporate boards of directors, state licensing boards, institutional review boards, consultants, and trainers. It may be appropriate for society ethics/review committees to adopt such practices as well.

5.6 The Effectiveness of Society Activities

As the public increasingly demands greater accountability on the part of the scientific community and as societies seek effective ways to promote research integrity, their activities must be subject to rigorous evaluation.

All codes of ethics of scientific and engineering societies encourage general good conduct. The codes encourage society to conduct and perform research honestly (including giving expert consultation and in delivering service). The work should be performed by working within the boundaries of competence, by following all applicable regulations and procedures). There should be no harm done to the discipline, to research subjects, to institutions, to clients, to the public, and to society). However, there are differences between the codes of ethics of the various societies, such as the level of specificity (i.e., articulated more abstractly as principles or as detailed expected behaviors), and the implied purpose (i.e., primarily to educate, to sanction, or to protect the public).

However, many scientific societies realize that the adoption of a code of ethics can be an important, but insufficient step for fostering responsible research practices. In seeking ways to reinforce the message carried by their codes, societies may engage in a range of activities. Furthermore, the range of activities reflects, at least in part, the fact that the societies are highly heterogeneous and some activities are a more appropriate fit to a specific society than others.

As the public increasingly demands greater accountability on the part of the scientific community and as societies seek effective ways to promote research integrity, these activities must be subject to rigorous evaluation. But neither resources nor strategies in support of evaluation appear to be a priority among the societies. The survey results revealed few means by which societies determine the effectiveness of their activities. Three indicated they conduct surveys and two mentioned informal feedback. Other categories mentioned once included outcomes of research projects, attendance at programs, meeting evaluations, annual reviews, peer review of research articles, disciplinary procedures, compliance with guidelines of society's instructions for authors, and the practice of addressing specific ethical concerns on a case-by-case basis.

Most, if not all, societies recognize that the following activities appear to be most effective for promoting research integrity:

1. publications on research ethics,
2. programs at annual meetings,
3. columns/articles in professional journals and newsletters, and
4. resource material with which mandatory compliance is specified, mentoring, and oversight of journal article reviewers.

Ethics committees, resource materials, and posting materials on a Website (unless a focal point of the site) were reported as least effective. But none of these appears to have been evaluated with any rigor. Indeed, it is not even clear what would constitute the criterion of "effectiveness" in order to draw valid conclusions. The reality is that these responses are more reflective of seat-of-the-pants judgments than any empirical evidence.

Scientific societies and professional associations should work closely together in developing and implementing codes of ethics as a way to bridge gaps in the understanding of ethical responsibilities across disciplines and professions. More research is needed on the importance of the societies (and other forces in the research system) in shaping the ethical climate in which scientists and engineers work. Worth explanation is how the exercise of professional discretion by individual scientists and engineers is affected by standards prescribed by his or her society.

In planning a research project, a clear delineation of roles, working relationships, credit allocation, and intellectual property policies is desirable. The design of methods of dispute resolution may help to promote responsible research practices and support collegial models for conducting collaborative research. Societies should consider adopting partnering agreements, conflict resolution mechanisms, and mentoring strategies in support of scientists and engineers, and students in the respective disciplines.

At present, there has been very little formal evaluation of the effectiveness of the society initiatives described in this report. More rigorous evaluation is essential if resources are to be efficiently allocated and if scientists and engineers, and the larger public are to have confidence in the self-regulatory functions of the societies. Such evaluation should be sensitive to the heterogeneity of the population of scientific societies.

Beyond impressionistic observations, little is known about the role and influence of scientific societies on research conduct.

There can be little argument with the notion that societies can play a key role in developing initiatives to help prevent ethical infractions and promote responsible research conduct. Yet, a scientific or engineering society may not always be a sufficiently impartial judge of allegations of research misconduct. Like all institutions, societies can overtly or subtly engage in cover-ups to protect their good name or to avoid possible litigation. Nevertheless, scientific societies can and should do more to promote research integrity.

In their role as publishers, societies have the opportunity to influence research conduct. Societies should review their codes of ethics to determine whether they appropriately cover publication ethics, which is a critical element in promoting research integrity. The society's leadership should work closely with new editors and new generations of researcher-scholars regarding ethical standards and their crucial role in helping to ensure the integrity of research.

Society journals should develop educational programs regarding publication policies that promote integrity in publishing scholarly work. In fact, scientific societies should establish a consortium of journal editors to develop, where appropriate, consistent standards for publishing scientific research. Furthermore, scientific societies should work together to establish a uniform policy regarding authorship in the context of multi-disciplinary research collaborations.

Criteria for authorship and the responsibilities – including relative contributions – of authors should be clearly stated by society journals. Specific standards for online publication should be developed by the societies.

Not enough attention is paid to the information and communications field, and only its positive effects and possible contribution to national development were mentioned while neglecting side effects and responsibilities.

In order for scientific research to be executed according to the ethical codes and be socially responsible, voluntary practice within the scientific community is more necessary than regulations from outside. Moreover, it is unconventional to solely depend on exterior regulations to deal with various social and ethical conflicts that arise during the research process. It is an individual scientist's/engineer's duty to execute socially responsible and ethical research practices. However, promotion of such acts, criticism of wrongful deeds, and enforcement of appropriate regulations should spring from the understanding of the entire scientific community. In other words, it is not a problem of the personal conscience of an individual scientist/engineer, but of the scientific community's firm understanding of the broader socio-cultural context.

Research misconduct must be examined comprehensively, in any narrow sense, by a society Board of Directors to construct a basis for strict penalties and in a broad sense, to encourage respectable research practices. Vagueness of the philosophical boundary of research misconduct does not undermine the concept of research misconduct, but instead demonstrates the intimate connection, not contradictions, between freedom of research and misconduct regulations. By relating "freedom of research" with exterior and interior regulations of the scientific community concerning research practices, including misconduct, one could encourage a positive attitude on the part of scientists and engineers toward the code of conduct for scientists and engineers.

The range of provisions for research integrity is subject to change according to circumstances and history, and even though research integrity promotion and misconduct prevention do not coincide, misconduct prevention should serve as a presupposition to research integrity promotion. However, the "regulatory approach," which assumes the acceleration of research integrity by intensifying regulations, does not correspond to the current reality; yet, it is irresponsible and unfair to simply reinforce regulations without any institutional measures that respect the view of the majority of scientists and engineers. Furthermore, reducing misconduct does not result in the revitalization of research integrity, and considering the collective structure of modern scientific research, it is not fair to criticize the conscience of the hapless field researcher without corresponding improvements to the research environment.

Researchers' responsibility and duty are not fulfilled only by a researcher's individual self-awareness and effort, but require an overall change in the atmosphere and structure of the scientific community or society as a whole. Hoping that in the age of science technology, a new plan to induce both individual responsibility and duty and responsible practice by the scientific community, as well as the participation by society at large would be included in a Code of Ethics: First, with the changing scientific research environment, researchers should first acknowledge the danger factor, an innate characteristic of scientific research, identify and estimate possible dangers, and diligently manage them. Second, researchers should try to remain unaffected by financial profit-loss calculations. Third, in order to prevent scientific misconduct, a strictly controlled research process and scientists and engineers' honesty are called for.

Fourth, scientific researchers can fulfill their responsibility by concisely and clearly explaining his research.

The invisibility of the individual in group-focused situations is real and individual scientists and engineers should be recognized as individual moral agents. In addition, since the only reaction available for an individual is to blow the whistle or resign, a firm protection system should be developed.

However, confusion may always arise due to the overlap between research ethics and ethics of science and engineering focusing the government's sole attention on research ethics problematic. Nevertheless, scientific research misconduct is a problem that cannot be regulated by general means but is a matter to be decided by scientists and engineers. A mature community, it should hold its own set of rules, and these rules should be brought forward and invoked for educational purposes. The role of scientists and engineers and engineers is very important in this problem-solving process, and therefore they should be actively engaged in the discussion. There needs to be positive ethics to which the entire scientific community could assent as well as fluid communication between scientists and engineers and the public. The current attitude of "let scientists and engineers deal with science and engineering problems," clearly demonstrates the exclusiveness of the scientific community and it is becoming more distant with the continuing specialization of scientific activities. The scientific community should initiate open communication with the public.

A *code of ethics* may mean that civil society dictates the conduct of scientists and engineers, yet scientists and engineers seem to respond indifferently towards it. However, scientists and engineers should conform to the common goal of society and moreover, create new values for civil society and try to arrive at a consensus in our society which eagerly pursues economic development.

A role of ethical-minded teachers, research supervisors or professors is to train scientists and engineers properly from the beginning, in school and in university (Chapter 3), to prevent further serious problems. Despite possible resistance due to the indifferent and disapproving view of scientists and engineers, who complain of more duties imposed, a code of ethics should be established and written in detail above the general level of conception. There are also issues related to reeducation of senior researchers and part-time researcher training. Regarding whistle blowing, the need for a consulting desk rather than anonymous reporting was needed to reduce the fear of consequences.

Many scientists and engineers are too occupied with their research to study or learn about research ethics on their own. Therefore, education is the most urgent issue.

Since there is not a clear distinction between misconduct and proper research, scientists and engineers should actively project ethical values and judge based on their common understanding as a group. That is the way to secure freedom of research. In this common understanding social values should be projected, and in reverse values respected in the scientific community should be diffused in society. This interaction between values and society is crucial, and it should be reflected in a Code of Ethics.

Scientific societies diverge in the roles they play regarding the promotion of ethical conduct among members of their disciplines.

There are at least three functions for scientific societies that are important to track and trace over time and across fields:

1. general education and professional development,
2. prevention and advisement, and
3. complaint handling and enforcement of codes of ethics.

While scientific societies vary in their levels of engagement in these three functions, they differ especially as to whether they are engaged in regulation of scientists and engineers within their disciplines.

These three functions are realized through a variety of specific activities including, but not limited to:

1. Production of code of ethics and other normative statements
2. Providing leadership internal to field of science and engineering (e.g., with departments)
3. Collaborating across fields of science and engineering and education
4. Providing leadership external to field (e.g., national science and engineering policy)

Scientific societies vary both in the activities they pursue and in their levels of effort. Also, activities can be high or low profile, symbolic or concrete. In addition, they can be implemented on a case-by-case basis or be part of a more systemic effort to address integrity and misconduct.

Studies of the actual practices of societies to encourage responsible research conduct and to avert misconduct are important. Research should include attention not just to the types of activities scientific societies pursue but also to the intensity of such efforts, the level of deliberativeness, and changes over time. Also, it is important to examine the indirect and direct effects as well as short-term and long-term influences of such activities on shaping professional knowledge, attitudes, and behavior.

5.7 Academic Freedom

Although not really a society, many universities develop their own culture and own laws so that they appear to outsiders to be a society and law unto themselves. Therefore, because there are references to academic freedom elsewhere in this book, it is appropriate at this point to delve briefly into the realms of academia and the meaning of the term academic freedom.

The academic tradition emphasizes, "intellectual honesty and critical self-discipline with respect to:

1. the scholarship of discovery;
2. the scholarship of integration;
3. the scholarship of application; and
4. the scholarships of teaching" (Hamilton, 2002, page 42).

Furthermore, academic freedom has been defined as "a condition of work, designed to enable academics without suffering adverse consequences in their employment" (Tight, 1988, page 4). This allows for expanding the current horizons of knowledge. Academic freedom also exists in the ethical space between, "the autonomous pursuit of understanding and the specific historical, institutional and political realities that limit such pursuits" (Scott, 1996, 177). Such freedom allows researchers to uncover, discover, contradictions, discrepancies, and information that has not been formerly revealed (Robinson, 2001).

However, the nature and status of a university depends on the extent to which academic staff appreciates, understands and behaves in an ethical fashion while enjoying their academic freedom (Steneck, 1984). Furthermore, external pressures that force universities to be more competitive in the expanding marketplace can be and have been, "corrupting both of the spirit of the university and academic freedom" (O'Hear, 1988, page 16).

In addition, there is the thought that the accumulation of knowledge has been due to academic freedom but this is only partly true. One must not forget the accumulation of knowledge that occurs outside of academia in governmental organizations and other non-academic (commercial) organizations. The contribution to knowledge of such companies as: ExxonMobil (USA), Eastman Kodak (USA), RIM (Canada), IMP (Mexico), IFP (France), BASF (Germany), Statoil (Norway), SASOL (South Africa), and the now-defunct Imperial Chemical Industries in Britain (as well as many other companies) cannot, and must not, be overlooked or underestimated.

In some instances and in a different realm of their operations, mainly because of the autonomy that they have been allowed, including the lack of a well-defined peer review system and the overall lack of accountability of the professors, universities may knowingly or unknowingly engage in unethical practices (Swazey et al., 1993). Issues of ethics generally occur on the boundaries of academic freedom and therefore raise questions about the need for discussion and consensus about the limits of academic freedom and, by extension, whether or not there should be limits to autonomy bestowed upon universities (Neave, 1988, page 39).

These issues have to be addressed within the notion relating to the definition of a university and focus on views of university functions, such as the development of critical thinking and participation in and improvement of the quality of life while promoting self-reflection (Metz, 2009, pages 179–80).

The modern university is an institution for teaching, learning, protection of the culture, contributor to economic growth and a knowledge factory which is a shift from the university as, "a simple community of scholars and students united by a search for a deeper understanding of nature of nature and humankind" (Pocklington and Tupper, 2002, page 5). Moreover, the university has become, "a series of specialized factions, disciplines, students and research activities united only by occupancy of a common territory..... factions though, independent, broker deals with each other, undertake research that the public does not understand and utilize a language that the public cannot understand" (Pocklington and Tupper, 2002, page 4–6), while professors establish academic tribes and territories.

In such a context, academic freedom is synonymous with academic subjectivity as individuals utilize disciplinary jargon to justify their actions and guard their respective territories. The university has also been viewed as radical when in fact, it is most conservative in its institutional conduct," (Kerr, 2001, page 71). It is also seen as, "a law unto itself; the external reality is that it is governed by history," (Kerr, 2001, page 71).

To mitigate these issues (recalling that the prime mandate of a university is to teach and foster leaning in the students) universities today have to adjust in three major areas:

1. growth,
2. shifting academic emphases, and
3. participation in the life of the wider society (Kerr, 2001, page 81).

This requires that universities contribute to the creation of an environment that explores both a more complete understanding of education, and a culture (and practice) that take education to higher levels of ethics and morals.

Furthermore, since academic freedom is, "socially engineered spaces in which parties engaged in specific pursuits enjoy protection from parties who would otherwise naturally seek to interfere in those pursuits," (Menand, 1996, page 3) the accountability for such freedom has to be persistently monitored, which becomes conducive to self-regulation within the university.

In an era of increasing demands for accountability, universities must make an ethical commitment to justify their claims for institutional autonomy and academic freedom not only to those within their walls but also to those outside.

As a result of the inclination to defer to academic authority in earlier times (Haskell, 1996, page 55), those with academic authority were simultaneously obligated to preserve their integrity and disciplinary recognition.

Academic freedom evolved through several phases. In the early years, such freedom was constrained because of a combination of financial, political, moral and religious concerns. Intellectual exchanges were only supposed to occur between competent academics who would clarify differences between error and incompetence (Hamilton, 2002, 20). Academic freedom is rendered special because of self examination by the faculty in peer review (Hamilton, 2002, 21). However, one must ask if the peer review system within a university actually exists as a formal means of evaluating the performance of all of the faculty members as well as the review of all academic treatises prior to publication.

To many, both inside and outside of the university system, the concept of academic freedom, implies opportunities to choose what topics one wants to investigate and how far one wants to go in that regard. Choice involves acting on and sorting out whatever one wants by examining the consequences of each choice, which requires making decisions about means and ends (Stehr, 2008, 28). In the determination of means and ends, ethical factors must be considered.

Statements about academic ethics, as reflected in the Codes of Ethics of disciplinary bodies usually establish parameters to guide the actions of professors but, in general, a faculty member (especially a faculty member at the top of the professorial rank) is really free if he is the one who decides on courses of actions. This means that the professor is free to present any material (objectionable or not) he chooses to students in whatever manner he wishes.

This is where responsibility and accountability and such accountability must be manifested in the behavior of the professor, which is related to ethical conduct. Every academic becomes obligated or it is the duty of academics to provide undergraduate and graduate students with certain assurances of ethical and moral behavior, hence, accountability, even though the concept of academic freedom implies that there are no boundaries to thoughts, words, and deeds as stated before. In fact this might be at least one reason why the credibility of academic institutions is being questioned, especially when individuality in academia begins to override the requirements of sociality and ethical behavior (Downing, 2005).

In summary, academic freedom means that a faculty member has the autonomy to teach, to perform research, and publish the results of that research but (what is often failure to recognize) within the boundaries of ethical and moral behavior. Students are good imitators of professorial behavior. What students see, students do.

Indeed, the mere act of engaging in unethical practices (which is not always covert) is also evidence of the fact that scientists and engineers (in academia or outside of academia) are not always rational (Chapter 9), although they may be able to rationalize their emotions.

Academic freedom must be used in an ethically acceptable fashion in teaching or research or both. Following a Code of Ethics is much more needed by an academic than an intellectual because the latter knows that he has the freedom of choice to produce, visualize and justify new ideas.

It is, however, the man by which this freedom of choice (i.e., academic freedom) is followed and practiced.

References

AAAS. 2000. *The Role and Activities of Scientific Societies in Promoting Research Integrity. A Report of a Conference.* American Association for the Advancement of Science. Washington, DC. April 10.

Caelleigh, A.S. 2003. Roles *For Scientific Societies In Promoting Integrity In Publication Ethics.* Science and Engineering Ethics, 9(2): 221–224.

Chalk, R., Frankel, M.S., and Chafer, S.B. 1980. *Professional Ethics Activities in the Scientific and Engineering Societies*. American Association for the Advancement of Science, Washington, DC.

Chalk, R. 2005. AAAS *Professional Ethics Project: Professional Ethics Activities in the Scientific And Engineering Societies*. American Association for the Advancement of Science, Washington DC.

Downing, D.B. 2005. *The Knowledge Contract: Politics and Paradigms in the Academic Workplace*. University of Nebraska Press, Lincoln, Nebraska.

Hamilton, N. 2002. *Academic Ethics*. Praeger Press, Greenwood Publishing Group, Santa Barbara, California.

Haskell, T.L. 1996. "Justifying the Rights of Academic Freedom in the Era of 'Power Knowledge." *In The Future of Academic Freedom*. L. Menand (Editor). University of Chicago, Chicago, Illinois. Page 43–90.

Iverson, M., Frankel, M.S., and Siang, S. 2003. "Scientific Societies And Research Integrity: What Are They Doing And How Well Are They Doing It?." *Science and Engineering Ethics*, 9(2): 141–158.

Jones, N.L. 2007. "A Code of Ethics for the Life Sciences." *Science and Engineering Ethics*, 13(1): 25–43.

Kerr, C. 2001. *The Uses of the University*. Harvard University Press, Cambridge, Massachusetts.

Levine, F.J., and Iutcovich, J.M. 2003. *Challenges in Studying the Effects of Scientific Societies on Research Integrity*. Science and Engineering Ethics, 9: 257–268.

Menand, L. 1996. *The Limits of Academic Freedom. In The Future of Academic Freedom*. L. Menand (Editor). University of Chicago, Chicago, Illinois. Page 3–20.

Metz, T. 2009. "The Final Ends of Higher Education in Light of an African Moral Theory." *Journal of Philosophy of Education*, 43(2): 179–201.

Murray, E.G.D. 1947. "The Value of Scientific Societies." *Canadian Journal of Comparative Medicine*, XI(2): 47–52.

Neave, G. 1988. "On Being Economical with University Autonomy: Being an Account of the Prospective Joys of a Written Constitution." *In Academic Freedom and Responsibility*. M. Tight (Editor). Open University Press, McGraw-Hill Education, Maidenhead, Berkshire, United Kingdom. Page 31–48.

O'Hear, A. 1988. "Academic Freedom and the University." *In Academic Freedom and Responsibility*. M. Tight (editor). Open University Press, McGraw-Hill Education, Maidenhead, Berkshire, United Kingdom. Page 6–16.

OSTP. 1999. *Proposed Federal Policy on Research Misconduct to Protect the Integrity of the Research Record*. Office of Science and Technology Policy, Executive Office of the President. Federal Register, 64(198): 55722–55725.

Pocklington, T. and Tupper, A. 2002. *No Place to Learn: Why Universities Aren't Working*. UBC Press, Toronto, Ontario, Canada.

Robinson, K.A. 2001. *Michel Foucault and the Freedom of Thought*. Edwin Mellen Press, New York.

Scott, J. 1996. *Academic Freedom as Ethical Practice. In The Future of Academic Freedom*. L. Menand (Editor). The University of Chicago Press, Chicago, Illinois. Page 163–180.

Stehr, N. 2008. *Knowledge and Democracy*. Transaction Publishers, Piscataway, New Jersey.

Steneck, N.H. 1984. Commentary: *The University and Research Ethics. In Science, Technology, & Human Values*. Fall 1984, 9(4): 6.

Swazey, J.P., Anderson, M.S., Lewis, K.S. 1993. "Ethical Problems in Academic Research." *American Scientist*, 81: 542.

Tight, M. 1988. *Editorial Introduction. In Academic Freedom and Responsibility*. M. Tight (Editor). Open University Press, McGraw-Hill Education, Maidenhead, Berkshire, United Kingdom. Page 1–5.

6

Codes of Ethics and Ethical Standards

6.1 Introduction

As best as is know at this time, every society and professional group has in place a range of norms to guide the behavior of its members. Similarly, colleges and universities are built on moral obligations, ethical responsibilities and principles and codes of behavior (Baca and Stein, 1983, page 7). Furthermore, there is a direct correlation between levels of moral outrage expressed and the importance of what is expected (the *norm*, an indicator of professionalism) from ethical standards (Braxton and Bayer, 1999, page 3).

In the realm of higher education, norms specify the desired practices with respect to teaching, research and service. Without norms, faculty members would be free to follow their own unconstrained preferences in teaching and research. Norms also represent what is considered important by a group articulating how professional choices mesh with services (Braxton and Brayer, 2002, 4).

It might be argued (unsuccessfully one hopes) that it is difficult to establish unambiguous ethical standards in academia, and this leads to a range of judgment calls (Whicker and Kronenfeld, 1994, page 9). The nature of this challenge is shaped by factors such as information overload and competency, both of which impact on departmental cultures, individual academic roles and identities. Furthermore, there is a relationship between academic communities and the ideas they express (Becher and Trowler, 2001, page 23). Academic culture comprises disciplinary knowledge, growth, enquiry methods, and research outcomes.

Whether or not they are in academic world or the commercial world or the governmental world, most scientist and engineers believe that they are honest; capable of acting not from instinct, but rather from a reasoned set of rules that are defined under varios (relevant *codes of ethics*). Briefly, a code of ethics provides a framework for ethical judgment (the incentive to do the *right thing*) by a scientist or engineer (Martin and Schinzinger, 2005; Fleddermann, 2008); although there are thoughts (not necessarily agreeable) that many plausible-sounding rules for defining ethical conduct might be destructive to the aims of scientific enquiry (Woodward and Goodstein, 1996).

For most of history, the discussion of ethics was dominated first by superstition and later by religious doctrine, and thus largely resistant to reasoned examination. It is only in the last few centuries have ethics been rigorously pursued outside of religious doctrine. Currently, even those who hold strong religious convictions are now dependent upon arguments from secular ethics to resolve disagreements with people of different religious beliefs and cultures. Likewise, most religious doctrines now accept that their texts should be viewed critically as products, at least in part, of human cultures.

Alternatively, if a scientist or engineer does not consider a religious text as the first, last, and only word on ethics, then he is left to find another basis for ethics (Schwartz, 2001, 2003, 2005). To reduce the problem of interpretation and the prevalence of inherent prejudices, one needs to seek a universal basis that can transcend the boundaries of faith and culture.

Despite the capacity for rationality, scientist and engineers have several significant obstacles to overcome when considering ethics. Foremost, there is evolutional memory and behavior patterns which can lead scientists and engineers to value themselves first, the scientific or engineering community second, and colleague third, if they are given any value them at all. Such patterns of thought are often referred to as, "moral intuition or moral instinct," i.e. that which feels right is right (or ethical) (Sommer, 2001).

However, not every scientist or engineer has the same instincts about ethics and not all instincts appear to be equally valid. Indeed, it is easy for any scientist or engineer to criticize or condemn the value or prejudices of others and so free themselves from ethical issues. Indeed, it is very difficult for scientists and engineers to distance themselves from their own views, so that they can dispassionately search for prejudices among the beliefs and values others hold.

Likewise, it is important that ethics, whenever possible, avoid deferring to potentially prejudiced instincts. As rational beings, scientist and engineers are not supposed or required to be slaves to these instincts.

Scientific research offers many other satisfactions in addition to the exhilaration of discovery. Researchers have the opportunity to associate with colleagues who have made important contributions to human knowledge; with peers who think deeply and care passionately about subjects of common interest, and with students who can be counted on to challenge assumptions. With many important developments occurring in areas where disciplines overlap, scientists and engineers have many opportunities to work with different people, explore new fields, and broaden their expertise.

Researchers often have considerable freedom both in choosing what to investigate and in deciding how to organize their professional and personal lives. They are part of a community based on ideals of trust and freedom, where hard work and achievement are recognized as deserving the highest rewards. And their work can have a direct and immediate impact on society, which ensures that the public will have an interest in the findings and implications of research.

Research can entail frustrations and disappointments as well as satisfactions. An experiment may fail because of poor design, technical complications, or the sheer intractability of nature. A favored hypothesis may turn out to be incorrect after consuming months of effort. Colleagues may disagree over the validity of experimental data, the interpretation of results, or credit for work done. Difficulties such as these are virtually impossible to avoid in science and engineering. They can strain the composure of the beginning and senior scientist alike. Yet struggling with them can also be a spur to important progress. Scientific progress and changes in the relationship between science and engineering and society.

Individuals operate according to their own beliefs of what is considered moral and what is not. There must be some over-riding code of ethics for scientists and engineers. However there will always be those scientists and engineers whose code is very simple: self first, self last, and, if there is anything left, self again.

The role of a *code of ethics* is characterized by both descriptive and prescriptive aspects. One can choose to affirm or deny role responsibility. Particularly when the occupant of a position is a scientist or engineer, it might be expected that the requisite knowledge and skills demanded by in these esteemed positions would be sufficient to guarantee research integrity except in a few extraordinary cases.

There is a direct relationship between the health of a profession and the maintenance of ethical standards, in academia and industry (Craine, 2004). Central to this relationship is the society culture which varies within and across societies.

The ethical culture of a society is a combination of intended and unintended outcomes that emerge from each of the facets of society (Figure 1). The nature of the ethical environment depends on how these facets impact at the membership level. Society leaders are basically mandated to enforce policies, rules and regulations. The manner in which that is done depends on the administrative style of leaders of each particular society.

The various scientific and engineering disciplines are world-wide professional disciplines (Harris, 2004). The members of these disciplines collect factual data and the ensuing treatment of the data to discover new arenas of knowledge is universal. No one can foresee the tortuous path of scientific and engineering investigation and know where experimentation and observation and may lead. Then there is always the mode of data interpretation.

The pursuit of science and engineering requires freedom of thought and, in the academic sense, unrestricted communication. It is through the professionalism of the members of the scientific and engineering disciplines that world knowledge and technology advances. Yet there are continuous reports of unethical behavior in the form of data manipulation, cheating, and plagiarism at the highest levels of the disciplines. The causes are manifold whether it is the need to advance in one of the chosen disciplines or to compete successful for and obtain research funding.

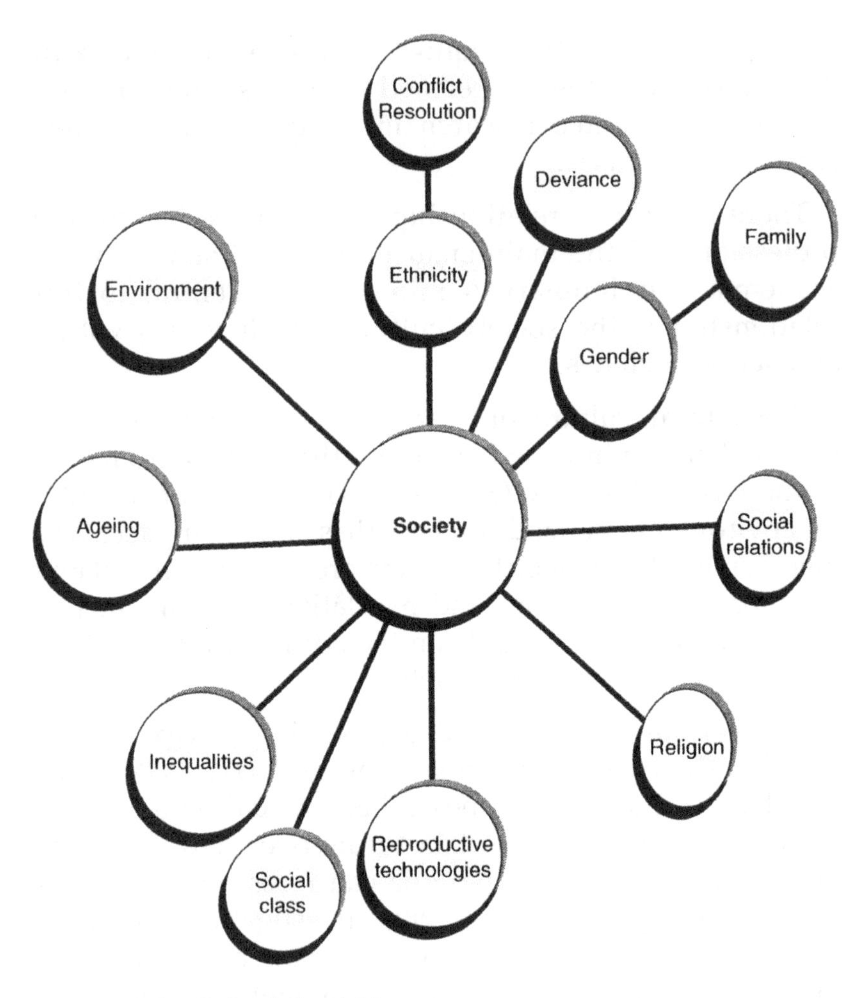

Figure 1. Ethically loaded social issues.

Research in the scientific and engineering disciplines offers the exhilaration of discovery. In addition, researchers have the opportunity to associate with:

1. colleagues who have made important contributions to human knowledge,
2. peers who think deeply and care passionately about subjects of common interest, and
3. students who can be counted on to challenge assumptions.

Also, scientists and engineers have many opportunities to work with different people to explore new fields and broaden their expertise, especially where disciplines overlap (Harris, 2004).

The reliability of scientific knowledge also derives partly from the interactions among scientists and engineers on an open and trustworthy basis (Davis, 1991; Alcorn, 2001; Altman, 1997). By engaging in such social interactions at society meetings and other forums where knowledge is presented and discussed, researchers must call on their technical understanding of the world and convince a collection (or community if the work is published in a technical journal) of peers of the correctness of their concepts, which requires a fine understanding of the methods, techniques, and conventions of technical research science and engineering.

However, research in any technical discipline can entail frustration and disappointment as well as satisfaction. Whether or not an experiment fails or a hypothesis turns out to be incorrect are all learning experiences. Instead of attempting to rationalize why an experiment failed, the investigator should determine if the experimental design was correct (or incorrect) or whether the collapse of a favored hypothesis is more likely the commencement of a modified hypothesis that is more logical than the previous hypothesis.

It is at this stage that many technical researchers decide that the experimental design was not incorrect or the failed hypothesis was not incorrect and they push forward to explain the experimental results. If the conduct of research in not monitored closely by peers and supervisors a situation exits where bending of the truth (it may not be called cheating but that is what it is) and the empirical objectivity of the researchers is lost. Often when this occurs, technical integrity has been forfeit.

For example, the experiment that failed becomes the experiment that succeeded because of a data point that has just been discovered. The defeated hypothesis becomes the successful hypothesis because the experimental design produced a datum point that the researcher was seeking. The means by which the datum point came about is another issue and is looked upon as good fortune by the supposedly unbiased and totally honorable involved researcher. Or the datum point was discovered in a blinding flash of untruthful inspiration by the researcher's co-worker who knew how important such a data point would be. The experiment that failed becomes the experiment that provided crucial proof of a concept.

On the other hand, too many points can be a hindrance to a researcher and lead to hours (or minutes or seconds) of heart rending consideration. The result might be that out of twenty four shotgun-patterned points on an x-y chart, eighteen points are omitted as flyers. The result is an x-y relationship on the chart that gives credence, even proof, to the hypothesis and results in wide acceptance of the hypothesis and copious honors for the researcher. After the success of such a brilliant hypothesis, there are few if any (perhaps because of funding constraints) who will repeat the work to determine if the data are correct. The hypothesis lives on and it is only after serious issues have been raised at some future time that the hypothesis is reworked. By then the original researcher may have retired after a distinguished career whose reputation in now beyond reproach. Younger researchers who could not make any sense of the hypothesis and report their data are at first criticized and ostracized.

Deleting data points is hardly uncommon, initially, all of the data points are printed on a scatter plot, and so-called flyers that do not match the plot are omitted. This is such a familiar situation in research that there are many reasons for deleting the non-conforming data.

This would imply that there are certain situations in which the practice is considered to be acceptable but such deletion actually amount to misrepresentation. Flyers can be influential or not influential insofar as they are far removed and inconsistent with the rest of the data, or are far removed but consistent with the rest of the data. In the former case, summarization and analysis of the data both with and without the outliers can be performed and the different inferences and conclusions are assessed, with and without the flyers. Nevertheless and in either case, all outliers must be reported. To do otherwise, is tantamount to technical fraud though dishonesty (intentional deception) (Resnik, 1998, 2000).

Another form of deception occurs when the reader is led to believe that the research design and execution were, according to the data points, flawless. Needless to say, both of these false impressions are intended not to further the ends of research but to further the self- interest of the investigator; such as making the publication more publishable, to garner honor or admiration or to discover a finding for the investigator and his research school.

Deception in data reporting is a remarkably reprehensible act. It dishonors scientists and engineers (from whom we expect the truth; anyone, scientist or not, can lie and deceive). It also dishonors the institution (which technically owns the data) so that the act of misrepresentation blemishes, by extension, the reputation of the institution; and to the extent that the data might someday be implemented in a commercials scenario.

Consequently, investigators who are blasé about deleting data points have probably neither thought through their moral obligations as scientists and engineers nor reasoned about the possible consequences their deception might influence future research participants. It is, as pointed out above, an expression of self-interest, and egomania.

In short, unethical behavior in the science and engineering disciplines is alive and continues to plague the minds of those who see such behavior as well as the general public who may experience such behavior when it is reported in the popular press. There is also the need to determine if ethics is alive. It is! But it is the minority (at least we hope a minority) of researchers who are the miscreants and give ethics a bad name because of their flaunting or bending of the truth or, for the want of a better word (if there is such a word) cheating.

6.2 Ethics

Ethics is not the same as feelings. Feelings provide important information for our ethical choices. Some people have highly developed habits that make them feel bad when they do something wrong, but many people feel good even though they are doing something wrong. And often our feelings will tell us it is uncomfortable to do the right thing if it is difficult.

On the other hand, ethics is not religion. Many people are not religious, but ethics applies to everyone. Most religions do advocate high ethical standards but sometimes do not address all the types of problems we face. Similarly, ethics does not mean following the law. A good system of law does incorporate many ethical standards, but law can be a function of power alone and designed to serve the interests of narrow groups.

Ethics does not involve following culturally accepted norms. Some cultures are quite ethical, but others become corrupt; the old adage when in Rome, do as the Romans do is not a satisfactory ethical standard.

Finally, ethics is not science or engineering. Science and engineering do not give direction on what is right and what is wrong. Scientific and engineering sub-disciplines may provide an explanation for human events but ethics provides reasons for how humans ought to act. Just because something is possible from a scientific or engineering aspect does not mean it is ethical.

The realm of ethics is concerned with standards and requirements for socially acceptable behavior in addition to following proper procedures for getting things done at any level of interaction: individual, group, organizational, community, governmental or regional.

Ethics has several strands:

1. descriptive ethics, that is, the actual behavior of scientist and engineers and the ethical requirements of their behavior;
2. normative ethics or identification of the values that sufficient to guide interaction;
3. meta-ethics which questions the meanings of all that ethics has been concerned about;
4. applied ethics or the application of normative rights to specific issues, disciplines and settings (Kitchener and Kitchener, 2009).

The requirements in this regard are stipulated in various Code of Ethics documents of scientific and engineering societies, such as the American Chemical Society (ACS) and the American Institute for Chemical Engineers (AIChE) as well as the many and other technical societies across the world. However, such codes do not resolve the issues which, in the final analysis, depend on personal decision-making and because knowledge claims must be free from bias, prejudice and personal values (Kitchener and Kitchener, 2009).

These codes cannot and must not be ignored by using claims of academic freedom. There are many instances where laws have been flaunted because an attorney has argued successfully that to obey the law is an infringement of his client's constitutional rights. It is not his client's constitutional right to bring harm to another person by violating a Code of Ethics.

There are several descriptions or definition of ethics including:

1. a system of moral principles, which are the ethics of human culture,
2. the rules of conduct recognized in respect to a particular class of human actions or a particular group; culture, which includes medical and Christian ethics;
3. moral principles, such as the ethics as of an individual which forbid betraying a confidence, and
4. the branch of philosophy that deals with values relating to human conduct; especially in respect of the correct or incorrect nature of certain actions and the motives behind such actions (Becker and Becker, 2002). Ethics is also "the normative science (and engineering) of conduct and conduct is a collective name for voluntary actions (Lillie, 2001)."

In this regard voluntary actions are those actions that could have been done differently (Lillie, 2001; Harris, 2004). Such actions may be good or bad, right or wrong, moral or immoral. Ethics focuses not on what men think but what they ought to think and do. An ethical science and engineering is an in-depth systematic study of the standards for judging right and wrong, good and bad, principles guiding means and how far we will or should go (Lillie, 2001; Howard and Çorver, 2008).

Whether the conduct of a scientist or engineer is correct or faulted may be (Lillie, 2001):

1. instinctive and discernible through individual actions,
2. intentional, which may be direct and motivating or indirect,

3. rooted in desire, which is a consciousness to act in a particular manner, or
4. a matter of calculated choice.

Indeed the actions of one person can impact on the actions of others and, as such, the general nature and direction of actions in a society may affect the choices of others and their level of consideration for moral standards (Lillie, 2001). This has an impact on concerns for the common good, levels of egoism and altruism and the eventual emergence of rights, duties and entitlements.

Ethics consists of those morally permissible standards of conduct each member of a group wants every other (member) to follow even if their following them would mean he or she had to follow them too. Thus, it is reasonable to assert that writing a properly functioning code of ethics is a collective task (Davis, 2007). Without a reasonable amount of group consensus concerning morally permissible standards of conduct relevant to the group, the code finds its home scribbled on a sheet of paper rather than in the actions and decisions of members of the group.

Ethical disagreements on rights, duties and entitlements are also possible and may take the form of disagreement in belief, when an individual believes in 'p' and another in 'not p' and as such one persistently challenges the other, and disagreement in attitude, when one has a favorable attitude and the other an unfavorable attitude towards an issue (Stevenson, 2006).

Furthermore, a code of ethics should fulfill many purposes within a scientific and engineering organization by:

1. increasing ethical sensitivity and judgment,
2. strengthening support for the moral courage, and
3. fine-tuning and the sense of identity of the organization.

Furthermore, there is a wide variety of codes of ethics, which are written by specific technical groups and which have their own purpose for existence and allow each group to face a set of ethical challenges that are unique to the group (Lichtenberg, 1996).

A code of ethics should be the benchmark of the acceptable standards of conduct, which members of a scientific or engineering organization make binding upon themselves. Often, codes of ethics prioritize commonly conflicting principles, which underlie the standards of conduct within an organization by prioritizing the principles in order to give guidance on how a member is to act as a responsible agent of the organization when situations require an element of compromise between principles (Davis and Stark, 2001). For example, as a profession, engineers have agreed that a commitment to public safety is essential when acting as a professional engineer. This agreement is reflected in professional codes such as any code of ethics for engineers. Likewise, codes of the scientific professionals should emphasize a similar priority to a commitment to upholding the safety and health of individuals. Yet the differences in the focus of their respective codes of ethics reflect the differences in the challenges that scientists and engineers face while attempting to address their respective concerns.

Because different groups are composed of different people with different purposes having differing means of accomplishing differing ends, priorities specific to one group may to be contradictory to those of another group. The reason for the differences in, say, priorities is because the tasks of one group, say engineers, may directly involve the improvement of conditions of society (or groups within society), whereas the priorities of another group may involve the improvement of the condition of individuals. In addition, the type of activities engaged in by members of an organization determines the situations in which the practice of ethical conduct may be jeopardized, and therein lies the reason for writing codes of ethics specific to an organization and the members.

This idea of moral responsibilities specific to a group is also central to the process of designing a code of ethics.

Generally, it seems that codes of ethics with a clearly defined purpose are more clearly stated and better organized. Many codes make effective use of defining a purpose by beginning the document with a preamble or a statement of intent. The preamble sets the tone of the document and outlines both the purpose of the organization and the purpose of the code. The statement of intent fulfills a similar purpose, but it focuses more on the purpose of the code and less on the purpose of the organization than does a preamble. Both are good ways to establish cohesion within the group that is essential to the proper functioning of a code of ethics.

To many, the code of ethics is merely a set of well meaning statements on a rarely seen and even less frequently and effectively implemented document but, in fact, the code of ethics must truly reflect the virtues of the group. Through a process of achieving consensus, writing a code of ethics becomes an excellent group-defining task. Consequently, a well-defined membership in the group, an outcome of devising and publicizing a code, aids in the functioning of the code. Through identification as a member of the group, a member's sense of duty to other members of the group and to the group's collective agreements expressed in the code is strengthened. As a result, the effectiveness of the code of ethics is also strengthened.

In addition, there are several items that must be considered when deciding what should be included in the code of ethics:

1. the persons, or groups of persons, affected by the organization or the members of the organization,

2. the main area of activity of the organization,
3. the unethical decisions and actions that the organization would like to prevent,
4. the means by which these the unethical decisions and actions can be prevented,
5. the types of ethical problems that members of the organization are most likely to encounter, and
6. the means by which conflicting principles be resolved (Davis and Stark, 2001).

The answers to these questions leads to the formulated what needs to be included in the organization's code of ethics, the next step is to decide what the code of ethics are for the organization. Just as principles within a code differ from group to group, so to, methods of organization differ from scientist to engineer, and wit in the respective sub-disciplines.

For example, the factors that may affect how an organization develops a code of ethics could include such aspects as:

1. the length of the code;
2. the means by which statements for inclusion in the code were formulated, and
3. the form of organization that is most familiar to the members (Schwartz, 2001, 2003, 2005). If relationships were a major consideration in the formulation of statements, it seems most appropriate to organize the code according to relationships. However, if relationships were not a major consideration but principles were a major consideration, it is most appropriate to organize the code according to principles and guidelines for the principles.

Thus, a code of ethics is a means of uniquely expressing the collective commitment of an organization to a specific set of standards of conduct while offering guidance in how to best follow those codes. As such, authors of a code of ethics should explore methods of organizing a code and use of language in the code that will be well received (and readable and understandable and not in *legalese*) by the membership. For example, William Shakespeare once stated, "kill all the lawyers," but this was taken out of context and was not the intent of that particular speech, as written in the play *Henry V*. Yet, many people are willing to take that statement out of context with their own individual preferences for the interpretation of the meaning. Codes of ethics should not allow the reader to do this! Giving guidance encourages the membership of an organization to develop and practice moral reasoning based on the collectively agreed-upon principles of the group enumerated in the code.

A workable code of ethics is written with the awareness that the code will be used in a variety of different situations, and each situation will prompt those involved to refer to the code for specific guidance (Harris, 2004). Thus, the code must be written with enough information to be of use in the specifics of a situation while remaining general enough to be used for a wide variety of situations. It is most likely this challenge that lies behind the inclusion of sections entitled such as, "Suggested Guidelines for use with the Code of Ethics, Standards of Practice or Rules and Procedures." In such sections, there are attempts by the organization to foresee situations one might encounter that call for ethical considerations. In many instances these guidelines attempt to provide guidance on how to resolve conflicting principles (Davis and Stark, 2001).

The brevity of many codes of ethics seems insufficient for fulfilling the many purposes of the codes. While codes that are short in length and content do illustrate an organization's commitment to fundamental principles, these codes may fail to give substantial guidance to the organization's members in situations which often require some sort of give and take between fundamental principles.

It is important for a code of ethics to include such guidance through the development of a code, because an organization makes collective agreements about what conduct is ethical and what conduct is unethical. In addition, the practice of ethics may be (some would insist always) situation-specific. A code of ethics lacking in guidance fails to address this very important aspect of the practice of ethics; thus, the code will likely fail at accomplishing its intended purposes.

Codes of ethics change with time due to changes in the organization, changes in society, and a desire by the organization management or by the membership to improve the effectiveness of a code. In this sense, a code of ethics should be thought of as a living document which must be adapted to the changing atmosphere of an organization, and the environment in which the organization operates. Through a process of revision, the codes of ethics keeps place with the times and changes in the law of the land.

From this perspective, the future of codes of ethics (and their ultimate usefulness) are left to the organization, to the membership, and the responsible fulfillment of the sections of the codes.

Since the actions of one person can impact on the actions of others and, as such, the general nature and direction of actions in a society may affect the choices of others and their level of consideration for moral standards (Lillie, 2001). This has a definite impact on concerns for the common good, levels of egoism and altruism and the eventual emergence of rights, duties and entitlements (Frankel, 1989).

Among scientists and engineers, ethical disagreements on rights, duties and entitlements are also possible and may take the form of disagreement in belief, when an individual believes in one aspect of the work and not the other; one persistently challenges his colleague, and disagreement in attitude, when one has a favorable attitude and the other an unfavorable attitude towards the data (Stevenson, 2006).

The extent and frequency of agreements and disagreements would vary with the extent to which there exists an ethical environment (Haydon, 2006). Schools and universities, like all other organizations, share an ethical environment. All societies have norms of conduct. Norms are synonymous with morals which signify how people should treat each other. Norm conformity is recognized as an obligation or duty and, in the absence of norms being identified, people can be guided by the consequences of their actions.

Values, laws and religious teachings are part of the ethical environment which must be evaluated and changed, if necessary (Haydon, 2006). This can happen through individual action, legal changes, education. Implicit in the creation and maintenance of an ethical environment is the emergence of regimes of reason or unreason; which are constitutive of conscious and unconscious, opposing and accepted values that often clash with each other in a society (Leitch, 1992).

An assessment of rights, duties and entitlements is also a moral issue and human moral capacities and judgments are shaped by personality, socialization, situational demographic (such as age, gender, and ethnicity) and broader societal factors.

For example, a PhD student following an experimental program decides that his original project title and synopsis require rigorous and taxing laboratory work, which may be beyond his capabilities. Although giving the supervisor/mentor glowing reports of the work as it (supposedly) progressed (but refusing to turn over the laboratory notebook for examination, and each time using some convenient excuse) has changed the program. He has been encouraged to do this by working with others who were not formally involved in the program and without the knowledge of the supervisor/mentor. No formal (or informal) requests were ever submitted by the student to formalize the change of plan and the supervisor/mentor discovers the deception at the time of drafting the thesis. When confronted with this issue, the student is unrepentant and the university powers-that-be are perceived to agree with the student's actions (insofar as the student received no form of reprimand). By allowing this, the university is encouraging the student to move into area of cheating and unethical behavior; and the word gets around that students can graduate by doing whatsoever they wish, without any form of guidance; essentially by flaunting the rules, or bending the rules to accomplish graduation.

Generally, such actions are due to the need to achieve a purpose, or to satisfy an interest or desire (Furrow, 2005). These factors do not impact institutions independently of each other but in combination. Indeed, morally inappropriate behavior is driven by thoughts and feelings that were cultivated and reinforced across time and space. Furthermore, moral autonomy is not achievable when personal desires, emotions, and inclinations persistently influence the judgment of a scientist or engineer. Moral autonomy must be exercised within certain ethical boundaries – even if it conflicts with individual's needs and desires and such needs and desires must be evaluated (Furrow, 2005). Reasoning is instrumental in helping to pursue and attain certain goals.

If the act performed by the individual scientist or engineer is not in his power not to perform, then he is responsible for that act and must face the consequences (Chisholm, 2008). This would establish the morality of the action given that to act morally is to act autonomously, not as a result of technical or social processes (Williams, 2006). However the selective orientation to autonomous or independent individual-level action is not to be shaped and reshaped on a whim. It must be reinforced by accountability and evaluation standards. In addition to these, the promotion of ethical behavior would serve to reduce ethical lapses in the academic environment (where responsibility is often taken lightly but authority reigns supreme) as well as in other environments (Kezar et al., 2008).

However, once a promise or commitment is made, scientist and engineers are obligated to keep it and such obligations are very difficult to escape (Furrow, 2005). Some scientists and engineers may not keep their obligations because they are not quite comfortable with themselves and/or because of others giving them different advice. The result is diminished willpower or intention to fulfill an obligation. Intentions are the outcomes of deliberating with oneself to decide what to do (Williams, 2006).

The assessment of rights, duties and entitlements is also a moral issue, which can be shaped by personality, socialization, situational demographic (age, gender, and ethnicity) and broader societal factors. Generally, scientist and engineers want to achieve a purpose or satisfy an interest or desire (Furrow, 2005). These factors do not impact independently of each other but in combination. Indeed, morally appropriate behavior is driven by thoughts and feelings that are cultivated and reinforced throughout generations. The argument is that moral autonomy is not achievable when personal desires, emotions, and inclinations persistently influence the judgment of scientists and engineers. However, moral autonomy has to be exercised within certain societal boundaries even if it conflicts with an individual's needs. In this regard, it is necessary for the scientist and engineer to evaluate their thoughts and desires (Davis and Stark, 2001; Furrow, 2005).

The reality of individual, group, organizational and cultural differences universally has generated a diversity of moral codes; most people do not subscribe to a single moral code. This has resulted in moral relativism which does not mean that there is no true objective moral code. Relativism has been justified on the basis of physical and cultural differences and the consequent promotion of tolerance for different views (Rachels, 2000). In the context of social changes, communication and interactions with other countries, there has been significant cross-fertilization of ideas and influence; orienting people to make judgments on levels of morality (Furrow, 2005).

It is generally known that once a promise or commitment is made, it very difficult to withdraw from the obligation (Furrow, 2005). Some scientists and engineers may not keep their obligations because they are not quite comfortable with themselves and/or because of others giving them different advice. The result is diminished willpower or intention to fulfill the obligation. Intentions are defined as, the outcomes of deliberating with self to decide what to do (Williams, 2006).

While it is true that beliefs are not under voluntary control, it is also true that scientists and engineers choose what to believe, and, as a result, choice is under our control. In this regard, it is essential for the scientist and enginery to remain open-minded and always be ready to evaluate arguments, findings, and the different perspectives of each person involved.

Consequently, it is necessary, in fact essential, to realize that:

1. the end does not justify the means;
2. a rational basis must be established for dealing with uncertainty in any type of research;
3. while researchers prefer to minimize errors, the outcome of such preferences must be thoroughly evaluated (Shrader-Frechette, 1994).

If the act that the individual performs is in his power not to perform, then he is responsible for that act and must face the consequences (Chisholm, 2008). This would establish the morality of the action given; that to act morally is, 'to act autonomously, not as a result of social processes'

(Williams, 2006). It must be noted, however, that the orientation to autonomous or independent individual-level action is shaped and reshaped by a changing society. As a result, the central influencing factor is the quality of individual-level socialization despite the changing nature of the context. It is further reinforced by law enforcement, cultural influences, accountability arrangements and monitoring and evaluation standards. In addition to these, the promotion of equity initiatives would serve to reduce ethical lapses in universities and other settings (Kipnis, 1983; Kezar et al., 2008).

6.3 Codes of Ethics

Codes of Ethics are intended to legally reinforce the need for respect for scientific and engineering data as well as for all other human beings independent of what anybody thinks about location, upbringing, gender, ethnicity, religious affiliation, age, culture, level of education, and other characteristics. Ethical issues have come and will remain at the fore because of a scientist or engineer's prioritization of differences as he seeks to arraign a more privileged position in his respective group, organization, and/or the world of academia. This requirement can be further compounded by procedural inconsistencies in any research project and the absence of a philosophical basis for discussions of ethics dictates the need for a more comprehensive theory to guide future research (Kitchener and Kitchener, 2009). This should focus on:

1. behavior and basic moral requirements;
2. ethical rules for decision making,
3. ethical principles that are used to justify ethical rules;
4. ethical theory providing explanations of how a scientist or engineer should act, and
5. meta-ethics which discusses and evaluates the meaning of ethics.

In addition, there seems to be much truth in the post-modern view of scientific and engineering research ethics that every research activity, question and decision has ethical underpinnings. Ethical issues must (they usually do) focus on:

1. research procedures of developing a title, research design, data collection, data interpretation and analysis, report writing, and communication of findings;
2. power relations of the researcher and researched;
3. views of respondents about future use of research findings, and
4. the researcher's assessment of his beliefs and values (Thomas, 2009).

Professionalism entails a multiplicity of tasks and a variety of new roles; not all individuals occupying these roles of trust have been adequately prepared for and socialized to them. Society is characterized by autonomous spheres of endeavor within which only some roles are realized, and therefore accountability may be weak or lacking. Conversely, actions are often collective, i.e., via team approaches to problem posing and problem solving, which can undermine individual responsibility. Indeed, the importance of recognizing the role of the society in contributing to incidences of research misconduct was noted during conference discussions. All of these potentially conflicting factors may make it difficult for a researcher to know with confidence what is ethically expected of him or her.

Briefly, research misconduct is, "fabrication, i.e., making up results and recording or reporting them, falsification, i.e., manipulation of research materials, equipment, or processes, or changing or omitting data or results such that the research is not accurately represented in the research record, and plagiarism, i.e., the appropriation of another person's ideas, processes results, or words without giving appropriate credit" (OSTP, 1999).

Codes of ethics are often considered to be controversial documents and some scientists and engineers even consider them to be unnecessary. On the other hand, others believe that codes are useful and important, but disagree (or are uncertain) about why codes are necessary.

Many scientific societies have developed codes of ethics that encompass a broad range of behavior and practice as a means of fostering research integrity. These codes presumably represent the ideals and core values of a profession, and can be used to transmit those values and more detailed ethical prescriptions as part of the education of scientists and engineers. They also provide standards for reviewing claims of misconduct and for sanctioning improper behavior.

When misconduct allegations are reviewed by societies, the results may not be made public, thereby diminishing the potential deterrent effect. Societies should, therefore, consider making public the outcomes of their misconduct review.

One of the pivotal questions faced by a scientific society is whether to institute measures to enforce its code of ethics with disciplinary proceedings and sanctions. Many societies choose not to engage in enforcement, using their ethics codes primarily for educational purposes. For other societies, ethics codes enforcement allows them to demonstrate their willingness to hold their members accountable for their conduct. Yet another option adopted by some societies is referral of a grievance to the institution that owns the data to conduct an investigation, with the society reserving the right to publicize the findings of that investigation.

The potential for and the limitations of codes of ethics to ensure research integrity provoke varying points of view. While codes are intended to codify standards of behavior in professional roles, their limitations are such that conduct cannot be guaranteed and, in some instances, cannot be predicted. The contexts of scientific research can present unique circumstances that create difficulty in describing behavior that is uniformly right or wrong. Any decision or dilemma requires an examination of competing values as well as good judgment and common sense, and the individual value systems of each member must also be factored into decision-making.

Therefore, the adoption of a code of ethics is significant for the professionalization of the members of a society because it is one of the external hallmarks testifying to the claim that the group recognizes an obligation to society that (hopefully) transcends mere economic self-interest.

Codes of Ethics shape the behavior of scientists and engineers and offer the means by which research should proceed.

Conceptual work needs to focus not only on potential determinants of research integrity and misconduct but also on the specific indicators of research integrity and misconduct. It is important that research examines positive ethical practice as well as research misconduct. In considering misconduct, intent is also important because the very same manifestations may happen by design, by inattention or inadvertence, or even out of ignorance. Furthermore, departures from ethical standards may be isolated events or part of a more general pattern or practice of research misconduct.

Under the broad umbrella of research integrity and misconduct, a program of study would want to include attention to all aspects of the research process from data collection through dissemination. For example, research could usefully address such elements of conduct as:

- Authenticity of the work process
- Fabrication of data

- Falsification of data
- Authenticity of work product
- Plagiarism
- Misappropriation of other's data
- Accurate reporting of results
- Having or using appropriate expertise in the conduct of research
- Authorship and appropriate credit
- Data access or sharing
- Protection of human subjects/animals
- Honoring agreements of privacy and confidentiality

Manifestations of research misconduct can often be very low profile, invisible activities. At times, they are modest transgressions that become large in their significance because they are incremental over time. How apparent or observable misconduct is and even when or where it takes place can be highly dependent on the research process itself. As suggested above, factors like the site of research (e.g., laboratory versus field) or even the mode of conducting studies (e.g., solo investigator versus multi-investigator team) can affect both the norms of and opportunities for misconduct.

If the provisions specified in a code of ethics can be supported with good reasons, there is no reason why a profession does not include an affirmation of those provisions as part of what it professes. This does not preclude individual members from autonomously accepting those provisions and jointly committing themselves to their support. In fact, there is always a strong positive case for professional codes of ethics. For those who disagree, perhaps an examination of the individual's ethics might be in order?

The code is to protect each professional from certain pressures (for example, the pressure to cut corners in laboratory work to get the next publication out. In fact, having a code of ethics allows a scientist or engineer to object to pressure to produce substandard work not merely as an ordinary moral agent, but *as a professional*. Scientists and engineers should be able to state, without recrimination, that as a professional, "I cannot ethically put personal or business concerns ahead of my professional ethics."

Supporting a professional code will help assure each scientist and engineer a working environment in which it will be easier than it would otherwise be to resist pressure to do much that the engineers would rather not do. Scientists and engineers should support the codes of their respective professions code; supporting the code helps make the profession a practice of merit, since the code should generate benefits for all scientists and engineers (David, 1991).

The possible functions of a code of ethics include: (1) a collective recognition by members of a profession of its responsibilities; (2) an environment in which ethical behavior is the norm and, therefore, expected; (3) the code can serve as a guide or reminder in specific situations; (4) the process of developing and modifying a code of ethics can be valuable for a profession; (5) a code can serve as an educational tool, providing a focal point for discussion in classes and professional meetings, and (6) a code can indicate to others that the profession is seriously concerned with responsible, professional conduct (Harris et al., 1995).

6.4 The Premise Behind Codes of Ethics

A code of ethics is adopted by a society or by an organization in an attempt to assist the society membership when called upon to make a decision (usually most, if not all) understand the difference between correct actions *and incorrect actions* and to apply this understanding to their decision (Annas, 2006).

Codes of ethics should be developed by all scientific disciplines; with the process of development offering ample opportunity for contributions from all sectors of a society's membership.

Ethics and publication standards are not always effectively transmitted from one generation of scientists and engineers to the next, or even to current members of a society. Hence, any effort to develop standards should be linked to a plan for their dissemination and for the education of those to whom they (will) apply. For example, ethics consulting services sponsored by societies may help members assess options for responsible conduct.

If a society decides to enforce its standards with review and disciplinary procedures, it should be prepared to devote adequate resources to do so effectively. Enforcement procedures should accord due process and ways to initiate a grievance should be commonly known.

Thus, for the scientist and engineer, a code of ethics often focuses issues related to work, although the code may also focus on social issues, and set out general principles about the beliefs of the society or organization on matters such as:

1. mission statement,
2. quality of work,
3. standards of behavior towards others,
4. privacy, and/or
5. the environment.

More important, the code should delineate proper procedures to determine whether a violation of the code of ethics has occurred and, if so, what remedies should be imposed (Luegenbiehl, 1983; Johnson, 1991; Ladd, 1991).

The effectiveness of such codes of ethics depends on the extent to which management supports them with sanctions and rewards. Violations of the code of ethics of a society or organization usually can subject the perpetrator or violator to the prescribed consequences, such as expulsion from the society or dismissal from the organization.

In some cases, a code of ethics may be adopted by a profession or by a governmental or non-governmental organization as a code of practice: which also regulates the behavior of the members of that profession. A code of practice may also be styled as a code of professional responsibility; which will discuss issues that need to be discussed and the difficult decisions that will often need to be made, and provide a clear account of what behavior is considered ethical or correct in the circumstances. In a membership context, failure to comply with a code of practice generally results in (or, really should result in) expulsion from the professional organization.

Codes of ethics are created in response to actual or anticipated ethical conflicts. Considered in a vacuum, many codes of ethics would be difficult to comprehend or interpret. It is only in the context of real life and real ethical ambiguity that the codes take on any meaning. In fact, the best way to use these codes is to apply them to a variety of situations and see what results. It is from the back and forth evaluation of the codes and the cases that thoughtful moral judgments can best arise.

The underlying premise of any code of ethics is that the scientist or engineer should not sacrifice professionalism by rejecting one or more of the guidelines in relevant code of ethics of the organization.

However, the relevant code of ethics should offer the following:

1. clear and unambiguous advice,
2. no opportunity for someone endorsing the opposite course of action also use the code to support his choice;

3. the different guidelines within the code should not give conflicting guidance but guidelines should point to the same outcome, and
4. the professional code of ethics should not conflict with the individual moral compass of the scientist and/or engineer (Davis and Stark, 2001).

Multiple factors shape the norms, values, knowledge, and conduct of scientists and engineers and thus should be part of any research agenda on research integrity. Since the research enterprise is itself a social process, there are a number of individual, situational, and structural influences that can affect what scientists and engineers believe and how they work in general as well as under special conditions.

There are a number of considerations for any scientific and engineering society regarding enforcement. Due process considerations are essential in a review of misconduct if expulsion from society membership is a possible outcome. In addition, reviewers of misconduct allegations must have the right to access all sources of relevant information. There should also be a plan for transmitting a finding of misconduct to appropriate persons/institutions should be in place to protect the integrity of the research record. All parties involved in the review of misconduct are vulnerable to being sued and junior scientists and engineers may be reluctant to participate in disciplinary proceedings out of fear of professional vulnerability.

Enforcement of a code of ethics is not an easy task and societies must be willing to expend sufficient resources to do it well. The question of whether enforcement will serve as a real deterrent to misconduct is by no means settled. Therefore, careful drafting or redrafting of society codes may permit enforcement while addressing some of these concerns.

6.5 Codes of Ethics and Peer Reviews

Basic research can provide an understanding of the manifestations of research integrity and misconduct and the factors that affect their occurrence. In addition, basic research can help refine measures of research integrity. This research base can also offer a solid framework for intervention strategies designed to have a positive impact on research behaviors. However, the research process does not end with the implementation of intervention activities. Evaluation of interventions is an important component of any agenda of inquiry.

As scientific societies become more intentionally involved in devising strategies to promote research integrity (or consider new approaches for doing so): they should carefully examine what they are doing and review the effects of any such actions.

Many scientists and engineers have long maintained that evaluation research should accompany planned organizational or institutional change whether in the public, private, or nonprofit sectors. Rigorous evaluation, especially when grounded in a commitment to continuous quality improvement; provides a framework for meaningful assessment and for self correction. As a research strategy, it permits examining the impact of interventions or actions through an assessment of both the implementation process and the outcomes for targeted groups. In many respects, it provides the link between theory and practice.

When scientific societies or, for that matter, academic institutions develop research integrity programs, evaluation research can play an important role in assessing the effectiveness of these initiatives. Furthermore, it offers additional empirical assessments of relationships among factors that are hypothesized to promote research integrity. Results from evaluation studies provide evidence about what works and does not work, which in turn contributes to program improvement efforts over time.

For example, if basic research finds that conforming to the standards of responsible research conduct is highly correlated with the level of knowledge people have about the standards, then this work has implications for the development of an intervention strategy. It would be logical to develop an educational program to help professionals learn what is considered appropriate versus inappropriate research behavior. But, what is an effective educational strategy? Is an on-line educational course going to be effective? Conversely, is an educational approach based on a mentoring model likely to be more effective? As with any intervention, often it is useful to conduct pilot projects where different strategies can be introduced, ideally on a randomized basis, and compared to determine the most effective methods.

Evaluation research provides the methodology for conducting such research, involving both a process and outcome evaluation of the interventions implemented. The key elements to consider in conducting an evaluation include:

1. defining program goals for a specified target audience (e.g., graduate students will be made aware of the ethical standards for research and the strategies for adhering to these standards),
2. designing and implementing activities to achieve these goals (e.g., an educational program consisting of a one-credit course established as a graduation requirement or developing an independent learning CD-Rom training module);
3. delineating in advance a plan for evaluation, which addresses issues of a) measurements and instrumentation (e.g., measures of knowledge that use a paper/pencil test or measures of decision-making that use hypothetical case scenarios); b) timing of data collection (e.g., at the end of each course); c) methods of analysis (e.g., quantitative); d) plan and format for reporting

the results; and e) implications for an organization's activities, and
4. reporting results and ensuring a system for linking knowledge gained through research to further organizational planning and action.

Evaluation research offers considerable benefits to scientific societies or other institutions seeking to introduce change. While the goals may be clear, the strategies likely to achieve those goals may be quite uncertain. Introducing change on an experimental basis with appropriate evaluation has the advantage of encouraging an organization to be open to change without a long-term commitment to any given strategy. Such a model generates evidence, which becomes valuable input into decisions about future changes that may need to be made to improve further the outcomes of programmatic efforts.

While evaluation methodology is useful when concrete activities are pursued, there are limitations to its implementation; when scientific societies are engaged in work that is more symbolic than programmatic At present, scientific societies have been limited in developing intentional interventions. Indeed, as we have discussed above, efforts within organizations are generally quite intermittent or weak (as distinct from unimportant). The strength of the activity or intervention needs to be of sufficient substance to warrant systematic evaluation and to have outcomes that can be specified.

In an effort to go mitigate unethical behavior, the "ethics review process" should be detailed in the code, although if a charge is brought against a member, where appropriate, it is recommended that the academic or other institution that employs the member should make the investigation and resolve the issue. When it is determined that an ethical violation has occurred, a recommendation is made to the society president for action the president must be able to follow specific guidelines. A finding of plagiarism may result in a letter of reprimand and an author can be barred from publishing in any society for up to five years; an author's correction or retraction should also be required. The penalties for fabrication or falsification need to be more severe. Publication of a retraction is mandatory and various publications, leadership roles, privileges and rewards are precluded. The society may decide to publish the charges and findings in the relevant society publications (e.g., a newsletter or weekly/monthly magazine). A report of the actions should also be forwarded to the author's employing institution as well as to the appropriate government offices if federal funds are involved.

In addition, the society must also be prepared to review and, if necessary, revise its code of ethics over a three-year period, even if the revised code is longer and more detailed than The Original Code.

Therefore, while intentional change of any scope should be evaluated, it is also the case that basic research reviewed by peers.

The development of the science and engineering disciplines has paralleled the reconciliation of ethical issues. In the process, concerns and mechanisms for accountability in research must be an imperative.

References

Alcorn, P.A. 2001. *Practical Ethics for a Technological World*. Prentice Hall, Upper Saddle River, New Jersey.

Altman, E. 1997. "Scientific Research Misconduct." In Research Misconduct: *Issues, Implications and Strategies*. E. Altman and P. Hernon (Editors). Ablex Publishing Corporation, New York.

Annas, J. 2006. "Virtue Ethics." *In The Oxford Handbook of Ethical Theory*. D. Copp (Editor). Oxford University Press, Oxford, England.

Baca, M.C., and Stein, R.H. Editors) 1983. "The Social Contract Nature of Academic Freedom." *In Ethical Principles Practices and Problems in Higher Education. Springfield* : Charles Thomas Publisher, Springfield, Illinois. Page 23–36.

Becher, T., and Trowler, P.R. 2001. *Academic Tribes and Territories 2nd Edition*. Open University Press, McGraw-Hill Education, Maidenhead, Berkshire, United Kingdom.

Becker, L.C., and Becker, C.B. (Editors). 2002. *Encyclopedia of Ethics 2nd Edition*. Routledge, New York. Volumes 1, 2, and 3.

Braxton, J.M. and Bayer, A.E. 1999. *Faculty Misconduct in Collegiate Teaching*. John Hopkins University Press, Baltimore, Maryland.

Chisholm, R. M. 2008. "Libertarianism: The Case for Free Will and its Incompatibility with Determinism." *In Reason and Responsibility*. J. Feinberg and R. Shafer-Landau (Editors). Thomson-Wadsworth, Belmont, California.

Crane, A.G. 2004. A *Challenge for the New Century*. The Rotarian, Evanston, Illinois. 182(7): 24.

Davis, M. 1991. *Thinking like and Engineer; The Place of a Code of Ethics in the Practice of a Profession*. Philosophy and Public Affairs, 20(2): 150–167.

Davis, M., and Stark, A. 2001. "Conflict of Interest in the Professions". *Vol. VI The Oxford Series*. Oxford University Press, Oxford, England.

Davis, M. 2007. "Eighteen Rules for Writing a Code of Ethics." *Science and Engineering Ethics*, 13(2): 171–189.

Fleddermann, C.B. 2008. *Engineering Ethics 3rd Edition*. Pearson Prentice Hall, Upper Saddle River, New Jersey.

Frankel, M.S. 1989. "Professional Codes: Why, How and With What Impact?" *Journal of Business Ethics*. 8: 109–115.

Furrow, D. 2005. *Ethics: Key Concepts in Philosophy*. Continuum Press, New York.

Harris, C.E., Jr., Pritchard, M.S., and Rabins, M.J. 1995. *Engineering Ethics: Concepts and Cases*. Wadsworth Publishing, Belmont, Califorinia.

Harris, C.E. 2004. "Internationalizing Professional Codes in Engineering." *Science and Engineering Ethics*, 10: 503–521.

Haydon, G. 2006. *Education, Philosophy and the Ethical Environment*. Routledge, New York.

Howard, R.A., and Korver, C.D. 2008. *Ethics for the Real World*. Harvard Business Press, Cambridge, Massachusetts.

Johnson, D.G. (Editor). 1991. *Ethical Issues in Engineering*. Prentice-Hall, Englewood Cliffs, New Jersey.

Kezar, A., Glenn, W.J. Lester, J., and Nakamoto, J. 2008. "Examining Organizational Contextual Features that Affect Implementation of Equity Initiatives." *The Journal of Higher Education*, 79(2): 125–159.

Kipnis, K. 1983. "Evaluating Codes of Professional Ethics. In Profits and Professions." *Essays in Business and Professional Ethics*. W.L. Robinson, and M.S. Pritchard (Editors). Humana Press, Clifton, New Jersey.

Kitchener, K.S., and Kitchener, R.F. 2009. "Social Science Research Ethics Historical and Philosophical Issues." *In The Handbook of Social Research Ethics.* D.M. Mertens and P.E. Ginsberg (Editors). Sage Publications, Thousand Oaks, California.

Ladd, J. 1991. "The Quest for a Code of Professional Ethics: An Intellectual and Moral Confusion." *Ethical Issues in Engineering.* Deborah G. Johnson (Editor). Prentice-Hall, Englewood Cliffs, New Jersey. Page 130–136.

Lichtenberg, J. 1996. "What are the Codes of Ethics for?." *In Codes of Ethics and the Professions.* M. Candy and S. Black (Editors). Melbourne University. Press, Melbourne, Australia.

Lillie, W. 2001. *An Introduction to Ethics.* Allied Publishers Limited, New Delhi, India.

Luegenbiehl, H.C. 1983. "Codes of Ethics and the Moral Education of Engineers." *Business and Professional Ethics Journal*, 2: 41–61.

Martin, M.W., and Schinzinger, R. 2005. *Ethics in Engineering 4th Edition.* McGraw Hill, New York.

OSTP. 1999. *Proposed Federal Policy on Research Misconduct to Protect the Integrity of the Research Record.* Office of Science and Technology Policy, Executive Office of the President. Federal Register, 64(198): 55722–55725.

Resnik, D.B. 1998. *The Ethics of Science: An Introduction.* Routledge Publishers, New York.

Resnik, D.B. 2000. "Statistics, Ethics, and Research: An Agenda for Education and Reform." *Accountability in Research.* 8: 163–188.

Schwartz, M.S. 2001. "The Nature of the Relationship Between Corporate Codes of Ethics and Behavior." *Journal of Business Ethics.* 32: 247–262.

Schwartz, M.S. 2003. "The Development of a Model Code for Ethics Professionals." *Professional Ethics.* 11: 3–16.

Schwartz, M. S. 2005. "Universal Moral Values for Corporate Code of Ethics." *Journal of Business Ethics*, 59: 27–44.

Shrader-Frechette, K. 1994. *Ethics of Scientific Research.* Rowman and Littlefield Publishers, New York.

Sommer, M.J. 2001. "Ethical Codes of Conduct and Organizational Context: A Study of the Relationship Between Codes of Conduct, Employee Behavior and Organizational Rules." *Journal of Business Ethics*, 30: 185–195.

Stevenson, C.L. 2006. The Nature of Ethical Disagreement. In Philosophical Horizons. S.M. Cahn and M. Ekert (Editors). Thomson-Wadsworth, Belmont, California. Page 284–288.

Thomas, V.G. 2009. "Critical Race Theory: Ethics and Dimensions of Diversity in Research." *In The Handbook of Social Research Ethics.*

D.M. Mertens and P.E. Ginsberg (Editors). Sage Publications, Thousand Oaks, California.

Whicker, M.L., and Kronenfeld, J.J. 1994. *Dealing with Ethical Dilemmas on Campus*. Sage Publications, Thousand Oaks, California.

Williams, B. 2006. *Ethics and the Limits of Philosophy*. Routledge, New York.

Woodward, J., and Goodstein, D. 1996. "Conduct, Misconduct, and the Structure of Science." *American Scientist*, 84: 481–490.

7

Integrity in Research

7.1 Introduction

Research programs have been used to transform concepts into theories and, simultaneous with this development, has been some degree of diffusion as researchers explore new lines of enquiry as they attempt to make their contributions to the literature (Smith, 2008, page 57).

Ethical issues permeate every stage of the research process from the provision of a title to the study of the analysis of the data as mentioned before (Reagan, 1971; NAS, 1992). There is a range of ethical issues emerging in the fields of qualitative and quantitative research. This has been and remains so for several reasons: quantitative research is rooted in rationality, objectivity and reflection can be used to correct/evaluate and logic of analyses done; and qualitative approaches to data collection are more personalized and allows for expressions of values, beliefs, motivations, emotions in sharing of information. In addition to the ethical responsibilities of researchers, respondents also have ethical responsibilities. And yet, researchers, for several reasons, may or may not adhere to their personal and/or professional ethics.

Science and engineering are built on a foundation of trust insofar as all scientific and engineering research results are an honest and accurate reflection of a researcher's work. Researchers equally trust that their colleagues have gathered data carefully, have used appropriate analytic and statistical techniques, have reported their results accurately, and have treated the work of other researchers with respect. When this trust is misplaced, integrity is called into question (Branscomb, 1985).

Furthermore, any scientist or engineer who is requested to be a coauthor should ignore the data in next-to-final draft before publication (after the data have been *massaged* to look presentable) and check the original data. If this is called into question and the professional standards of science are violated, researchers are not just personally affronted and they feel that the base of their profession has been undermined. This would impact the relationship between science and society (Bertozzi, 2009).

Research is based on the same ethical values that apply in everyday life, including honesty, fairness, objectivity, openness, trustworthiness, and respect for others. A *scientific standard* or an *engineering standard* refers to the application of these values in the context of research. Examples are openness in sharing research materials, fairness in reviewing grant proposals, respect for one's colleagues and students, and honesty in reporting research results.

The most serious violations of standards have come to be known as *scientific misconduct*. The government of the United States defines misconduct as, "fabrication, falsification, or plagiarism in proposing, performing, or reviewing research, or in reporting research results" (Bertozzi, 2009), and this should also include, "peddling hype or myths to the media" (Roy, 1999), which can lead to fame and/or notoriety as well as additional funding for future programs. All research institutions that receive federal funds must have policies and procedures in place to investigate and report research misconduct, and anyone who is aware of a potential act of misconduct must follow these policies and procedures.

Some scientists and engineers believe that the few highly publicized cases of research misconduct, generally manifested as falsification, fabrication, or plagiarism, are just the tip of the iceberg. Others suggest that the relatively few known cases indicate that the overall incidence of misconduct is low. However, these cases generally don't cover publication disputes, unless they involve plagiarism (Ritter, 2001).

Nevertheless, scientists and engineers who violate standards other than fabrication, falsification, or plagiarism are said to engage in questionable research practices. Scientists and their institutions should act to discourage questionable research practices through a broad range of formal and informal methods in the research environment. They should also accept responsibility for determining which questionable research practices are serious enough to warrant institutional penalties. Standards apply throughout the research enterprise, but scientific practices and engineering practices can vary among disciplines or laboratories. Understanding both the underlying standards and the differing practices in research is important to working successfully with others.

Therefore, integrity in research is the application of truth to all activities involved in research. Research consists in working to make new discoveries and expressing that work in the form of publication on the technical and/or the patent literature.

Furthermore, research integrity or research ethics has many facets, of which examples are:

1. defining research misconduct,
2. conducting and reporting experiments,
3. protecting research subjects,
4. giving and claiming credit, and finally
5. reporting misconduct (Whitbeck, 1998).

In fact, "integrity in research is about promoting excellence (high quality) in these activities, and this positive emphasis on excellence should be kept paramount in thinking about honesty in research" (Martin and Schinzinger, 2005).

There have been attempts to define misconduct in research using both wider and narrower definitions, developed in specific contexts, and for different purposes. For example, if the purpose is to punish wrongdoers, a narrow and legalistic definition is likely to be favored. On the other hand, if the purpose of the definition is to assure high-quality research, in all its dimensions, a wider definition might be adopted which will typically emphasize honesty in conducting and reporting experiments; while also including theft, other misuses of research funds, and sexual harassment among researchers (Martin and Schinzinger, 2005). Misconduct is misconduct and it is preferable that any form of misconduct be recognized, whatever forms the definition may take.

Research is systematic enquiry whose goal is communicable knowledge:

a. It's systematic because it is pursued according to a defined plan;
b. It's an enquiry because it seeks to find answers to questions;
c. It's goal-directed because the objects of the enquiry are posed by the task description;
d. It's knowledge-directed because the findings of the enquiry must go beyond providing mere information;

e. Nevertheless, it's communicable because the findings must be intelligible to, and located within some framework of understanding for, an appropriate audience.

Scientific and engineering research takes place in many settings, including universities, government labs, and corporations. The requirements vary somewhat, according to the applicable guidelines and regulations, but truthfulness and responsibility applies in all settings. Furthermore, the activity of reporting research is an important part of conducting research. Research results are useful when they are reported clearly, completely, in a timely manner, and honestly (Martin and Schinzinger, 2005).

The application of ethics to research activities seeks to ensure that research is conducted with acceptable standards of morality in order to preserve integrity, validity and reliability of the study (Ryan, 1995; Fleddermann, 2008). While standards for conducting research zero in on the study itself, ethical issues emphasize people. Such issues include: concerns about fraud, misconduct, harm to subjects, infringement of rights, conflicts of interest, and misrepresentation of self and others (Altman, 1997; Hernon and Calvert, 1997). This also includes the manipulation of the statistics (Huff, 1954; Gibilisco, 2004). Many government departments have adopted new codes of conduct for research performed by staff, consultants, and contactors (Heilprin, 2003; Hileman, 2005). Professional bodies have stipulated codes of conduct to guide scientific and engineering practices but this may not have been enough because of the general lack of (at least reported) disciplinary actions against any perpetrators of misconduct.

Furthermore, the ethical aspects of scientific and engineering research revolve around the responses to: (1) the ethically proper way to collect, analyze and report all aspects of a study, and (2) researcher-respondents interactions (Kitchener and Kitchener, 2009).

However, a deep commitment to scientific integrity is best achieved by providing sound training in scientific practices, the ethical conduct of science and engineering, and by creating institutional and professional environments that reinforce the high standards addressed in that training. Ideally, this educational process should begin early in the training of future scientists and engineers and continue through the most senior career stages.

Individual scientists and engineers, research institutions, and professional societies bear primary responsibility for the integrity of science and engineering; the legitimacy of scientific practices, and the investigation and response to cases of alleged research misconduct. Institutions and units within them that train and hire investigators are responsible for selecting, socializing, educating, supervising, and disciplining research scientists and engineers (Ryan, 1995).

In all cases, the philosophy behind the modern approach to scientific and engineering research is to:

1. be liberal about the sources of conjecture and hypothesis at the commencement of research,
2. be skeptical in the handling of data and argument, and
3. be astringent in testing findings and explanations on the completion of research.

Individual scientists and engineers, research institutions, and professional societies bear primary responsibility for the integrity of science and engineering, the legitimacy of scientific practices, and the investigation and response to cases of alleged research misconduct. Institutions (and units within them) that train and hire investigators are responsible for selecting, socializing, educating, supervising, and disciplining research scientists and engineers (Ryan, 1995).

This responsibility must be shared, however, by professional societies and the journals that review and publish results of research. Any activity of the Federal Government in this domain should support and complement the institutional role, and federal intervention should occur only when institutions fail to fulfill their responsibilities.

For the individual researcher, integrity embodies a range of good research practice and conduct and includes several facets (NRCNA, 2002):

1. This includes Intellectual honesty in proposing, performing, and reporting research;
2. While Maintaining accuracy in representing contributions to research proposals and reports;
3. It is fairness in peer review;
4. It is collegiality in scientific interactions (including communications and sharing of resources);
5. It's also transparency in conflicts of interest or potential conflicts of interest;
6. It embodies protection of human subjects in the conduct of research;
7. While encompassing humane care of animals in the conduct of research;
8. Furthermore, it's adherence to the mutual responsibilities between investigators and their research participants.

For an institution, integrity is a commitment to creating an environment that promotes responsible conduct by embracing standards of excellence, trustworthiness, and lawfulness.

The reliability of scientific knowledge also derives partly from the interactions among scientists and engineers on an open and trustworthy basis. By engaging in such social interactions at society meetings and other forums where knowledge is presented and discussed, researchers must call on their technical understanding of the world and convince a collection or community (if the work is published in a technical journal) of peers of the correctness of their concepts, which requires a fine understanding of the methods, techniques, and conventions of technical research science and engineering (Cassell, 1982).

It is at this stage that many technical researchers decide that the experimental design was not incorrect, the failed hypothesis was not incorrect, and they push forward to explain the experimental results. If the conduct of research is not monitored closely by peers and supervisors a situation exits where bending of the truth (it may not be called *cheating* but that is what it is) and the empirical objectivity of the researchers is lost. And when this occurs, technical integrity has been forfeited.

For example, the experiment that failed becomes the experiment that succeeds because of a data point that has just been discovered. The defeated hypothesis becomes the successful hypothesis because the experimental design produced a datum point that the researcher was seeking. The means by which the datum point came about is another issue and is looked upon as good fortune by the supposedly unbiased and totally honorable involved researcher. On the other hand, the datum point was discovered in a blinding flash of untruthful inspiration by the researcher's co-worker who knew how important such a data point would be. The experiment that failed becomes the experiment that provided the crucial proof of a concept.

Yet, too many points can be a hindrance to a researcher and lead to hours (or minutes or seconds) of heart rending consideration. The result might be that out of twenty four shotgun-patterned points on an x-y chart, eighteen points are omitted as flyers (outlying data points). The result is an x-y relationship on the chart that gives credence, even proof, to the hypothesis and results in wide acceptance of the hypothesis and copious honors for the researcher. After the success of such a brilliant hypothesis, there are few if any (perhaps because of funding constraints) who will repeat the work to determine if the data are correct.

The hypothesis lives on and it is only after serious issues have been raised at some future time that the hypothesis is reworked. By then the original researcher may have retired after a distinguished career whose reputation in now beyond reproach. Younger researchers who could not make any sense of the hypothesis and report their data are at first criticized and ostracized.

Flyers can be influential or not influential. In other words, they can be far removed and inconsistent with the rest of the data or be far removed but consistent with the rest of the data. In the former case, one can do summarization and analysis of the data both with and without the outliers because the inferences and conclusions are different with and without the outliers. In the latter case, separate analyses are similar and not a problem, andthe outliers have little effect on inferences and conclusions. Nevertheless and in either case, all outliers must be reported, to do otherwise is scientific fraud. Obviously, when data deletion changes the results of the study or misrepresents the study, the act of deletion is unethical (Resnik, 2000).

Deception in data reporting dishonors scientists and engineers (from whom the truth is expected). Consequently, investigators who nonchalantly delete data points have probably not thought through their moral obligations as scientists and engineers nor have they thought of the possible consequences their deception might someday wreak on research participants.

High ethical standards in research are keys to protecting clinical research data, ensuring the quality of our research, maximizing the benefits, and minimizing the risks of further development. High ethical standards are also essential for any researcher to obtain approval for his methods, data, medicines and for his peers to put their trust in his research program (and/or products).

The researcher should apply the same high standards wherever he operates, including contract organizations and researchers where collaboration is essential. Collaborators should use principles that are aligned with those of the researcher. In fact, for research that is conducted as part of a collaboration, the researchers raise awareness of his policies at the beginning of the collaboration and include clauses in any collaboration agreement (verbal or preferably written) requiring adherence to the same high standards of ethics.

All collaborators, including the original researcher, should continuously evaluate the risks and benefits of the program at every stage: from initial research, through it, beyond the development stages, and then after a new product is approved for manufacture or a new process is approved for development.

Most researchers are ethical and approach studies with the best interests of the sponsor. Few would argue for further restrictions, and most appreciate the extent of regulatory latitude that exists. However, with these freedoms come crucial decisions that researchers must address both pragmatically and ethically. Obviously, the need to hire skilled people and provide them with sufficient training and oversight to ensure a patient's welfare is of extreme importance. Doing this, however, requires significant investments of time, money, and patience. The benefits pay dividends in terms of quality of data collected from the project. If scientists and engineers make appropriate, ethical choices and responsibly delegate their research-related duties, everyone wins: the scientists and engineers, the sponsor, and most importantly, science and engineering. Unfortunately, unethical behavior in the science and engineering disciplines is alive and continues to plague the minds of those who see such behavior as well as the general public who may experience such behavior when it is reported in the popular press (Fleddermann, 2008).

There is also the need to determine if ethics is alive. It is! Yet, it is the minority of researchers who are the miscreants and give ethics a bad name because of their flaunting (bending of the truth) or, for the want of a better word cheating.

7.2 The Nature and Conduct of Research

Research is an activity enabling scientists and engineers to test some hypotheses or conclusions and contribute to knowledge (Shrader-Frechette, 1994). Research is also been defined as the process of making and proving claims (Altman, 1997). Research ethics informs researchers about how to conduct them when carrying out their studies and thereafter.

Research ethics regulations have traditionally focused on informed consent, breaches of confidentiality, stress, injury, coercion, invasion of privacy and deception. The ethical conduct of research protects participants from harm, but individual and/or private interests may intervene and thwart the attainment of public goals.

While research norms have been stipulated by various research councils and professional associations, government-mandated research regulations are absent, and, apply only to drugs investigations where they exist (Shrader-Frechette, 1994).

Thus, there is more than one way of defining research, and there are several traditions as to how research should be carried out.

Research is systematic enquiry whose goal is communicable knowledge:

1. It is systematic because it is pursued according to a defined plan;
2. It is an enquiry because it seeks to find answers to questions;
3. It is goal-directed because the objects of the enquiry are posed by the task description;

4. It is knowledge-directed because the findings of the enquiry must go beyond providing mere information;
5. Nevertheless, it's communicable because the findings must be intelligible to, and located within some framework of understanding for, an appropriate audience.

Whether or not researchers conduct scientific research, they have an implicit obligation to the society as a result of training and education that they had received (Shrader-Frechette, 1994, page 24). Most of the ethical issues arise with respect to methodological value judgments and such value judgments should be specified even if they are defensible (Shrader-Frechette, 1994, page 54–55).

Scientific results must also be presented in a manner that would avoid future misuse or misinterpretation. Membership in a profession carries with it an implicit commitment to pursue the welfare of the profession. This is partly done by avoiding hasty, unconfirmed statements, incomplete analyses, and by speaking out about these in the studies of peers; thus, the significance of peer reviews. This is why many journals have stipulations to deal with fraud and may require researchers to place their raw data in a special archive (Shrader-Frechette, 1994, page 57). However, different research applications often carry different degrees of risk for the public and, as such; researchers must aspire to high standards of reliability and validity in order to minimize damaging implications.

The philosophy behind the modern approach to scientific and engineering research is to:

1. be liberal about the sources of conjecture and hypothesis at the commencement of research,
2. be skeptical in the handling of data and argument, and

3. be astringent in testing findings and explanations on the completion of research.

Research misconduct is significant misbehavior that improperly appropriates the intellectual property (or contributions) of others, that intentionally impedes the progress of research, that risks corrupting the scientific record, or compromises the integrity of scientific practices. Such behaviors are unethical and unacceptable in proposing, conducting, reporting research, or in reviewing the proposals or research reports of others.

A deep commitment to scientific integrity is best achieved by providing sound training in scientific practices and the ethical conduct of science and engineering. Also by creating institutional and professional environments that reinforce the high standards addressed in that training. Ideally, this educational process should begin early in the training of future scientists and engineers and continue through the most senior career stages.

In scientific research, no formal process exists for reviewing questions about the scientific integrity of individuals and assessing and periodically renewing their professional membership and privileges in the scientific community. Thus, institutions bear particular responsibility for maintaining high professional standards (Ryan, 1995). In fact, there is a direct relationship between the health of the academic profession and the maintenance of ethical standards.

Central to this relationship is a departmental culture which varies within, and across campuses. Departmental cultures are characterized by:

1. the willingness of academics to act responsibly at all time,
2. the maintenance of self-regulation and peer review (within the boundaries of academic freedom and collegial self-governance),

3. exposure to the requirements of academia beyond the sub-discipline of a scientist or engineer, and
4. the willingness of academics to look out for each other.

Indeed, culture in many respects may be deemed to be more important than rules or regulations because it provides a means for dealing with tensions and pressures at all levels including interpersonal relations and professional relations.

In the modern world, many scientists and engineers are not committed to think of the consequences of their actions (Kearney, 1999), the focus is on personal image. Such occurrences render it possible for any mechanical expression of responsibility to be eroded. The scientist and engineer have to be responsible first before one can become or act like a professional and the demonstration of responsibility cannot be talked into being. Where there are interactional bonds, there is a commitment to be responsible.

Often when a professor has stolen an idea or concept and a complaint has been made to the university authorities, many of you have been told, "it is only a young professor seeking funding for his project," and the matter has been dismissed by the university management and the board of trustees (Board of Regents or Board of Trustees).

The academic tradition emphasizes intellectual honesty and critical self-discipline with respect to the scholarship of discovery; the scholarship of integration; the scholarship of application; and the scholarships of teaching. (Hamilton, 2002).

However, academic freedom is "a condition of work, designed to enable academics without suffering adverse consequences in their employment" (Tight 1988, 4). However, the integrity of the academic staff depends on how well they appreciate, understand and behave in an ethical fashion while enjoying their academic freedom.

In some instances and in a different realm of their operations, universities may engage in unethical practices because of the autonomy that they have been allowed. Issues of ethics generally occur on the boundaries of academic freedom; therefore, raising questions about the need for discussion and consensus about the limits of academic freedom (Neave, 1988) and, by extension, whether or not there should be limits to universities' autonomy.

The modern university is an institution for teaching, learning, protection of the culture, contributor to economic growth and a knowledge factory. The university was a community of scholars and students united by a search for a deeper understanding of nature and humankind. However it has now become a series of specialized factions, disciplines, students and research activities, united only by occupancy of a common territory (Pocklington and Topper, 2002). Professors establish academic tribes and territories in such a context that academic freedom is synonymous with academic subjectivity, where individuals utilize disciplinary jargon to justify their actions and guard their territories. The university has also been viewed as radical when, in fact, it is most conservative in its institutional conduct, and it is also seen as a law unto itself (Kerr, 2001).

The expectations for the responsible conduct of research are complex and not always well defined, leaving guidance for the responsible conduct of research disorganized. Some responsible practices are defined through law and institutional policies that must be followed. Others are set out in non-binding codes and guidelines that should be followed. Still other responsible practices are commonly accepted by most researchers but not written down. Instead, they are transmitted informally through mentoring, based on the understandings and values of each mentor. This situation is further complicated by the fact that researchers are not routinely tested on their knowledge of responsible practices or licensed. Moreover, their behavior as researchers is inconsistently monitored and the penalties for irresponsible behavior vary considerably.

Most researchers do care about responsible behavior in research and pay a great deal of attention to best research practices. The fact remains, however, that it can take some effort to find out what these practices are and how to act when the complex rules for responsible practice seem to conflict with one another.

Concern about misconduct in research first surfaced in the early 1980s following reports of cases of egregious misbehavior. One researcher republished under his own name dozens of articles previously published by others. Other researchers (in one way or another) falsified or fabricated research results. To make matters worse, it seemed as if research institutions sometimes ignored, or deliberately covered up problems, rather than investigate them. Eventually Congress stepped in requiring Federal funding agencies and research institutions to develop research misconduct policies.

Furthermore, even though Federal policies technically apply only to federally funded research, many research institutions apply Federal research misconduct policies to all research. Many research institutions have also broadened the basic Federal definitions to include other inappropriate practices. In combination, Federal and institutional research misconduct policies define research practices that researchers must avoid. Failure to do so can result in the termination of employment or ineligibility to receive Federal funding.

Research misconduct policies provide guidance on responsible conduct in three areas:

1. They establish definitions for misconduct in research;
2. They outline procedures for reporting and investigating misconduct;

3. They provide protection for whistleblowers (persons who report misconduct) and persons accused of misconduct.

Thus, the definitions of misconduct in research and the procedures for handling allegations of misconduct in research form the basis for effective self-regulation in research.

At first glance, all aspects of research conduct would appear to be governed. However, this is not actually the case as there are several vagaries that allow scientists and engineers the freedom in which to conduct their research.

The ultimate question for scientific and engineering researchers is reduced to the best way to juggle various aspects of the project; including, How to study protocol requirements, and how to handle financial and professional pressures imposed by the sponsor.

A Code of Ethics is much more needed by an academic than an intellectual because the intellectual knows that he/she has to produce, visualize, articulate, and justify new ideas, approaches and relationships without resorting to any unreasonable or questionable practices. An academic has to specify the terms and conditions of freedom while intellectuals naturally have a better understanding of freedom or thought. Because they believe ideas are not constrained by circumstances they can enjoy their freedom more. The Academics do not necessarily enjoy theirs and this may account for the lack of creativity in much of their work.

7.2.1 Single Investigators

As professionals, researchers have not been particularly concerned about rules for self-regulation. Since the goal of research is to advance knowledge through critical inquiry and scientific experimentation, it has commonly been assumed that the normal checking that goes on in testing new ideas is sufficient to keep researchers honest. Based on this assumption, research arguably does not need specific rules for self-regulation because it is, by definition, an activity that routinely monitors itself.

The lack of a perceived need for specific rules poses problems for researchers who want guidance on responsible research practices. Intellectually and professionally researchers organize their lives around fields of study. However, the societies that represent many fields of study for the most part have not developed comprehensive guidelines for responsible research practices. Many do have codes of ethics, but most codes of ethics are simply general statements about ideals and do not contain the specific guidance researchers need to work responsibly in complex research settings.

Fortunately, there are a few important exceptions to this last generalization. Comprehensive descriptions of responsible research practices can be found.

7.2.2 Team Investigators

The problems that can arise from a single investigator may be overcome by appointing a principal investigator to the project.

The principal investigator (PI) is charged with either conducting research activities of his own or supervising those who do. In reality, few scientist and engineers run their own studies. Instead, they hire qualified technicians or laboratory assistants to conduct basic experimental procedures, study the data, perform data assessments, and to keep accurate records of the laboratory activities.

In many cases, the regulations or conventions do not require that the principal investigator has any specific training or expertise, other than a current/former investigator in the laboratory with (hopefully) expertise in the area under study. Therefore, the extent to which study procedures are delegated and the level of experience, training, and education of those to whom the tasks are delegated are left to the discretion of the principal investigator.

In addition, principal investigators frequently designate sub-investigators. Most often, they designate sub-investigators from within the department where the research is being carried out (although there may be no requirement that the sub-investigators have specific training or expertise). These individuals are appointed to act as a surrogate for the principal investigator, and they work with the remainder of the project team just as the principal investigator would.

7.2.3 Misrepresenting Credentials

Misrepresenting credentials (lying on a resumé) is another, but common, type of deception. Researchers have been to forge credentials, which can be either blatant or take more subtle forms (Ogden, 1999; Martin and Schinzinger, 2005).

Some candidates lie on their resumés, some candidates embellish a little, while some embellish a lot, and others just lie. Most of the time, that lie is about their education. The key to stop such practices is to check resumés thoroughly, not just checking the address and telephone number, but by thoroughly checking every line item listed for education and employment. Without assiduously checking the facts and claims, it is impossible to determine who will include untruths on his resume.

The indications (but not conclusive signs) that a candidate may be lying on their resume are that the candidate is: well dressed, well spoken, well experienced in the field, on the defensive when asked to verify his education, and unable to produce evidence for his education.

Fake diplomas are also used to misrepresent an applicant's educational attainment. A search of the Internet will produce several online diploma mills that are willing to provide the custom replicated diplomas from any learning institution. The selling companies are part of a growing number of Internet sites where people can purchases unearned credentials from real universities.

One of the earliest cases discussed by the NSPE Board of Ethical Review (Case 79-5) was about an engineer who received a PhD from a nondescript (diploma mill) organization that required no attendance or study at its facilities. The engineer then listed the degree on all his professional correspondence and brochures and the NSPE board believed (or has ruled in the past) that when listing a PhD, there is no reason to identify the university from which the degree was obtained. Merely by listing the advanced degree alone, it is widely understood that it conveys an earned doctorate.

7.2.4 Misleading Listing of Authorship

Misleading listing of authorship is another area where deception can be perceived to occur and the order of authors' names in many scientific and engineering disciplines is usually understood to convey information about the relative contributions of the authors, with the earlier listing indicating greater contributions.

Authorship conventions may differ greatly among disciplines and among research groups. In some disciplines the group leader's name is always last, while in others it is always first. In some scientific fields, research supervisors' names rarely appear on papers, while in others the head of a research group is an author on almost every paper associated with the group. Some research groups and journals simply list authors alphabetically.

In some disciplines, the listing order is not considered important and alphabetical listing is the order of the day.

Many journals and professional societies have published policies (guidelines) that lay out the conventions for authorship in particular disciplines. These policies state that a person should be listed as the author of a paper only if that person made a direct and substantial intellectual contribution to the design of the research, the interpretation of the data, or the drafting of the paper, although students will find that scientific fields and specific journals vary in their policies. Just providing the laboratory space for a project or furnishing a sample used in the research is not sufficient to be included as an author, though such contributions may be recognized in a footnote or in a separate acknowledgments section. The acknowledgments sections can also be used to thank others who contributed to the work reported by the paper.

On the authors' side, a frank and open discussion of how these guidelines apply within a particular research project (as early in the research process as possible) can reduce later difficulties. Sometimes decisions about authorship cannot be made at the beginning of a project. In such cases, continuing discussion of the allocation of credit generally is preferable to making such decisions at the end of a project.

Decisions about authorship can be especially difficult in interdisciplinary collaborations or multi-group projects. Collaborators from different groups or scientific disciplines should be familiar with the conventions in all the fields involved in the collaboration. The best practice is for authorship criteria to be written down and shared among all collaborators.

Above all, it is unethical to omit the name of a coauthor that makes a significant contribution to the research (Chapter 8). It then becomes a question of the nature of the contribution by the proposed coauthor. Was the person a technician solely operating a spectrometer and doing nothing else? Conversely, was he a high level technician or a professional who operated the spectrometer and presented an interpretation to the other authors?

Answers to these and related question should prepare the way to clearly designate the authorship of the paper.

7.3 Collecting Research Data

Ethical issues permeate every stage of the research process from the provision of a title to the study onto the analysis of the data as mentioned before. There are a range of ethical issues emerging in the fields of qualitative and quantitative research. This has been and remains so for several reasons: First, quantitative research is rooted in rationality and objectivity and reflection can be used to correct/evaluate the logic of analyses done, and secondly qualitative approaches to data collection are more personalized and allow for expressions of values, beliefs, motivations, emotions in sharing of information.

In addition to the ethical responsibilities of researchers, respondents also have ethical responsibilities. More often than not respondents do not breach their ethical commitments, spoken or unspoken. Researchers, for several reasons, may or may not adhere to their personal and/or professional ethics.

Research ethics regulations have traditionally focused on informed consent, breaches of confidentiality, stress, injury, coercion, invasion of privacy and deception. The ethical conduct of research protects participants from harm and enlightens them on the goals of research. For example individual and/or private interests may intervene and thwart the attainment of public goals.

Whether or not researchers conduct scientific research, they have an implicit obligation to the society as a result of training and education that they had received (Shrader-Frechette, 1994, 24). Value-freedom in research is impossible because human beings cannot be completely objective with respect to the exact margin of error, choice of statistical test, sample selection, research designs, data interpretations, assumptions and theories. Most of the ethical issues arise with respect to methodological value judgments and such value judgments should be specified even if they are defensible (Shrader-Frechette, 1994, 54–55).

Because of the complexity of scientific and engineering research, mistakes and errors are inevitable (Bertozzi, 2009). Nevertheless, researchers have an obligation to the public, to their profession, and to themselves to be as accurate and as careful as possible. Scientific disciplines have developed methods and practices designed to minimize the possibility of mistakes, and failing to observe these methods violates the standards of science and engineering. Every result must be carefully prepared, submitted to the peer review process, and scrutinized even after publication.

Beyond honest errors are mistakes caused by negligence. Haste, carelessness, inattention (any of a number of faults) can lead to work that does not meet scientific standards or engineering standards. Researchers who are negligent are placing their reputation, the work of their colleagues, and the public's confidence in science at risk. Errors can do serious damage both within science and in the broader society that relies on scientific results.

Scientific and engineering data must also be presented in a manner that would avoid future misuse or misinterpretation. Membership in a profession carries with it an implicit commitment to pursue the welfare of the profession. This is partly done by avoiding hasty, unconfirmed statements, incomplete analyses and by speaking out about these in the studies of peers, thus the significance of peer reviews. This is why many journals have stipulations to deal with fraud and may require researchers to place their raw data in a special archive (Shrader-Frechette, 1994). However different research applications often carry different degrees of risk for the public and, as such, researchers must aspire to high standards of reliability and validity in order to minimize damaging implications. This raises concerns for epistemic and ethical objectivity.

Research on scientific misconduct has found that there are several categories of people who may engage in unethical practices, deliberately or not: new faculty members who have not been properly mentored, individuals seeking promotion or tenure, and those who like to see their name in print.

Organizational justice research has focused on processes that shape justice perceptions and evaluations. It has been established that motivations specify the desired conclusion (Blader and Bobocel, 2005). Many organizations support the importance of procedures for outcomes (Blader and Bobocel, 2005). Research has also unearthed several factors in addition to perceptions of fairness that impact on organizational justice (Gilliland and Paddock, 2005).

Some examples of ethical issues in research are:

1. failing to keep important analysis of documents of a period of time,
2. maintaining incomplete records of findings,
3. seeking the status of co-author without making a significant contribution to the article,
4. not allowing one's peers access to data collected and analyzed (especially after the article was published),
5. exploiting research assistants without acknowledging their assistants, and
6. bias in sampling (Barnbaum and Byron, 2001).

It is unethical for researchers to ignore the role of language in the making of meanings in the lives of the researched (Mertens et al., 2009). Researchers have also been identifying the biases in study findings which indicate that minority ethnic groups as being four times as likely as whites to be schizophrenic. They have criticized the unusually large correlations between race and social class and hurricane survival in Southern United States. This brings to the forefront the issue of social justice. Ethical issues also resonate in the choices and representations of dimensions of diversity to be researched (Mertens et al., 2009).

Ethical concerns also surface when looking at the criteria for: fairness, the study's ability to elicit from respondents information that they were unaware of, unawareness of social construction of reality by others.

Ethics, it has been established, is concerned with what should and should not be done and this is one of the requirements of a profession. Professional ethics constitute standards that are widely accepted within the profession (Schwartz, 2009). Generally the stipulations of ethical associations worldwide emphasize: high technical standards, a certain range of abilities, skills, and cultural knowledge, integrity, honesty, and respect, and responsibility. These are supposed to be borne in mind when developing, carrying out and reporting research results (Wolf et al., 2009).

Sometimes ethical concerns of researchers emanate from their awareness of the entities or communities or organizations that they represent, or from attempting to be neutral or from holding on to a specific set of principles In turn, the public's assessment of research or evaluation research in particular would focus on the approach to the study, degree of accuracy and reporting of results (Wolf et al., 2009). In an effort to balance clients' and societal needs evaluation researchers for example have to meander their way through ethical concerns and maintain ethical standards despite differences in stakeholders' interpretations. Juxtaposed in the realities of evaluation research are the 'change agents' role of evaluators, personal values of evaluators and the persistent need for objectivity in research. The outcome has been differences in the ethical orientations of individual researchers (Wolf et al., 2009).

With respect to experimental research on scientific and engineering issues (*randomized experiments*, sometimes called wildcat experiments), it has been established that their partial success in identifying cause-effect relationships is useful, bringing value to same; their role in decision-making and their contribution in reducing the cost of wrong decision-making must continue to be valued. Once this approach is providing the best possible answer in the circumstances, then it is doing what good ethics requires. Ethical concerns persist however with respect to risks and benefits and decision about which causal relation is more important to be investigated (Mark and Gamble, 2009).

7.3.1 Bias in Analytical Methods

Bias is a form of self-deception, which is sometimes motivated irrationality but other times it constitutes a more purposeful evasion. For example, researchers suspect an unpleasant reality, perhaps sensing that the data are going against what they want to believe. Then, instead of confronting the data honestly, they purposefully disregard the evidence or downplay its implications. The purpose and intention involved is typically unconscious or less than fully conscious (Martin and Schinzinger, 2005).

The accuracy of a test is a measure of how close the test result will be to the true value of the property being measured. As such the accuracy can be expressed as the *bias* between the test result and the true value. However, the absolute accuracy can only be established if the true value is known (Speight, 2002).

In the simplest sense, a convenient method to determine a relationship between two measured properties is to plot one against the other. Such an exercise will provide either a line fit of the points or a spread that may or may not be within the limits of experimental error. The data can then be used to determine the approximate accuracy of one or more points employed in the plot. For example, a point that lies outside the limits of experimental error (a flyer) will indicate an issue of accuracy with that test and the need for a repeat determination (Speight, 2002).

However, the graphical approach is not appropriate for finding the absolute accuracy between more than two properties. The well-established statistical technique of regression analysis is more pertinent to determining the accuracy of points derived from one property and any number of other properties. There are many instances in which relationships of this sort enable properties to be predicted from other measured properties with as good precision as they can be measured by a single test. It would be possible to examine in this way the relationships between all the specified properties of a product and to establish certain key properties from which the remainder could be predicted, but this would be a tedious task.

The example is the researcher who omits eighteen out of twenty four points on the basis that only six of the points were true and the remainder,the eighteen points that he omitted or discarded, were flyers.

This is bias in favor of the researcher's theory that he must prove to be the correct theory, for whatever reason.

The impact of analytical bias on scientific and engineering medical decisions is mostly unknown. A large margin of error may be acceptable in some circumstances, whereas other scenarios demand more accurate and precise laboratory measurements. Often, scientist and engineers interpret laboratory results within the larger context of the project history and physical examination, but the influence of imprecision in laboratory data on a scientist or engineer's assessment can be dangerous, if not fatal.

7.3.2 Misuse of the Data

Data misuse occurs when data obtained (through experimentation) is used in the wrong context and may even be data from another researcher that is used without the user's consent. The data can be used for support of an incorrect theory. Another example is when a scientist or engineer uses data that has been entrusted to them in a manner not intended by the owner of the data.

The related issue, *data protection,* is safeguarding data against misuse. Ways in which this is done is by keeping data under lock and key whether it is in a locked safe or on a computer hard-drive where it is protected by encryption, firewalls, and user authentication.

Such systems prevent any access without a key, combination, or password and will record the details (time, terminal, logged in ID) of both successful and unsuccessful access attempts. This provides traceability and so deters casual miss-use.

7.3.3 Falsification and Fabrication of the Data

Falsification of data is the selective alteration of data collected in the conduct of scientific investigation or the misrepresentation of uncertainty during analysis of the data. Falsification also includes the selective omission/deletion/suppression of conflicting data without scientific or statistical justification.

Falsification includes such practices as:

1. The alteration of data to render a modification of the variances in the data;
2. The entry of incorrect dates and experimental procedures in a laboratory notebook or in any other record keeping device,
3. The misrepresentation of the results from statistical analysis,
4. The misrepresentation of the methods of an experiment such as the equipment used to conduct the experiment,
5. The addition of false or misleading statements in the manuscript or published paper,
6. The publication of the same research results in multiple papers; this is self-plagiarism. This

includes presenting the same set of slides at a series of meetings in which only new one slide is added for each meeting that, literally, adds nothing to the presentation, but is included to seemingly add another conclusion and for the author to be invited to other meetings.

7. The providing false statements about the extent of a research study in an abstract submitted for publication and oral presentation at a professional society meeting.

Fabrication of data is the intentional act of creating records that do not exist and for which there is no basis in fact with the intent to mislead or deceive. In short, the data is a pipe dream or has been conjured up for various reasons, none of which are legitimate!

Researchers who manipulate (fabricate or falsify) their data in ways that deceive others, even if the manipulation seems insignificant at the time, are violating both the basic values and widely accepted professional standards of science and engineering. Researchers should draw conclusions based on their observations of nature. If data are altered to present a case that is stronger than the data warrant, the researchers mislead their colleagues and potentially impede progress in their field or research. They undermine their own authority and trustworthiness as researchers. And they introduce information into the scientific or engineering record that could cause harm to the broader society.

Because of the critical importance of methods, scientific and engineering papers must include a description of the procedures used to produce the data, sufficient to permit reviewers and readers of a scientific or engineering paper to evaluate not only the validity of the data but also the reliability of the methods used to derive those data. If this information is not available, other researchers may be less likely to accept the data and the conclusions drawn from them. They also may be unable to reproduce accurately the conditions under which the data were derived.

7.3.4 Plagiarism and Theft

Plagiarism is intentionally or negligently submitting the work of others as one's own. It is also claiming credit for someone else's ideas or work without acknowledging it, in contexts where one is morally required to acknowledge it (LaFollette, 1992).

Plagiarism is also the theft of intellectual property and is not unlike stealing from a commercial business. A special case of plagiarism is the, "frowned upon but not always unacceptable," practice of *self plagiarism* in which an author will use segments of his own published material (e.g., methods section of a scientific paper) in a new publication without reference.

Plagiarism and falsification of data or fabrication of data are the primary means of scientific fraud. Whether data are made up, copied from someone else, or manipulated to achieve some desired end result, it's always fraud. But perhaps the more interesting question concerning fraud is why it happens.

Scientists and engineers who believe that they deserve more recognition are more likely to falsify, plagiarize or manipulate the data in order to report successful results. This has been so since the era of Newton, Dalton, Darwin, and Freud as they sought fame and prestige. Small scale deviant practices are likely to persist because, despite the canons of scientific research scientists and engineers can always attribute small inconsistencies to unavoidable errors that accompany or infiltrate all research.

On the other hand, and quite often, the reason is money. There were several environmental labs in the 1980s and 1990s whose employees were caught changing the time clock on their GC/MS data systems or changing the baseline on a chromatographic analytical method (processes known colloquially as time traveling and peak shaving). In another case, data were shown to be completely fictitious. A lab received samples and sent out data with no intervening lab procedures. At the behest of the U.S. Environmental Protection Agency and state regulators, federal marshals swooped down on the lab, impounded its data, and took most of the staff to jail (Ryan, 2002).

One of the major determinants of judgments of the degree of responsibility is whether a controllable act is perceived or intentionally committed or due to negligence (Werner, 1995). Since judgment can only be reliably made after some period if observation or investigation. There is a general feeling that whether practices have increased.

Before deciding whether an ethical crisis exists, we have to determine whether one of three situations exists:

1. whether ethical standards are unknown and unclear,
2. whether they are clear but ignored, or
3. whether they are being followed. (McDowell, 2000).

Whether or not there is a crisis in professional responsibility depends very much on the extent to which individuals were responsible and disciplined before acquiring professional status. The fact remains that the search for truth, knowledge and understanding of the world pose powerful ethical demands for the individual who wants to be part of a community of individuals who call themselves scientists and engineers (Guba, 1990). Indeed, methodological, analytical and ethical issues are closely interconnected (Ryen, 2009) particularly so because we have to relate with people in doing research, people whose attitudes, values, perceptions of issues vary.

Whenever conflicts of interest interfere with the conduct of research, it should not be undertaken (Bok, 2006).

7.4 The Controls

One of the pivotal questions faced by a scientific society is whether to institute measures to enforce its code of ethics with disciplinary proceedings and sanctions. Many societies choose not to engage in enforcement, using their ethics codes primarily for educational purposes. For other societies, ethics code enforcement allows them to demonstrate their willingness to hold their members accountable for their conduct. Yet another option adopted by some societies is referral of a grievance to the institution that owns the data to conduct an investigation, with the society reserving the right to publicize the findings of that investigation.

Ideally, prevention of scientific misconduct is the best protection of the public as well as of the reputation of the various scientific disciplines. To develop an appropriate focus on ethics standards, one should consider how a scientific community functions. The behavioral messages of established faculty members, for instance, are a significant source of learning.

The influence of the hidden or informal curriculum may run counter to the educational messages of the formal means of communicating normative behavior and expectations. Based on studies, it is observed that trainees and junior colleagues model their professional behavior, to a large extent, on what their leaders do, not what they say. Established scientists and engineers are effective if they openly explain their difficult decisions as based on issues of right and wrong. In other words, modeling is a primary factor in assuring ethical conduct.

The most effective control is the development and publication of Codes of Ethics should be developed by all scientific disciplines, with the process of development offering ample opportunity for contributions from all sectors of a society's membership. However, ethics and publication standards are not always effectively transmitted from one generation of scientists and engineers to the next, or even to current members of a society. Hence, any effort to develop standards should be linked to a plan for their dissemination and for the education of those to whom they (will) apply. For example, ethics consulting services sponsored by societies may help members assess options for responsible conduct.

If a society decides to enforce its standards with review and disciplinary procedures, it should be prepared to devote adequate resources to do so effectively. Enforcement procedures should accord due process and ways to initiate a grievance should be commonly known.

When misconduct allegations are reviewed by societies, the results may not be made public, thereby diminishing the potential deterrent effect. Societies should, therefore, consider making public the outcome of any review of the misconduct by a member, no what his level in the scientific or engineering community.

In their role as publishers, societies have the opportunity to influence research conduct. Societies should review their codes of ethics to determine whether they appropriately cover publication ethics, a critical element in promoting research integrity. The society's leadership should work closely with new editors and new generations of researcher-scholars regarding ethical standards and their crucial role in helping to ensure the integrity of research. Society journals should develop educational programs regarding publication policies that promote integrity in publishing scholarly work.

The scientific societies should establish a consortium of journal editors to develop, where appropriate, consistent standards for publishing scientific research. Scientific societies should work together to establish a uniform policy regarding authorship in the context of multi-disciplinary research collaborations.

Criteria for authorship and the responsibilities, including relative contributions, of authors should be clearly stated by society journals. Furthermore, specific standards for online publication should be developed by the societies.

There should be no cover-up or attempted cover-up of misconduct in any of the on scientific or engineering disciplines.

Once misconduct by a member of any society has been proven, there should be no show of wrist-slapping. The member responsible for the misconduct should be expelled from the society and it made known publically why he is no longer welcome as a member of that society.

Furthermore, in order to keep one's nose clean, any scientist or engineer who is requested to be a coauthor should ignore the data in next-to-final draft before publication (after the data have been massaged to look presentable) and check the original data (Chapter 8). If there are inconsistencies in the transposition of the data from the laboratory notebook to the would-be draft for publication, the invited co-author should make noises to have this explained and, if necessary corrected, keep a paper trail as means of exoneration.

Finally, a checklist (which is not necessarily all-inclusive) is presented below that contains a range of questions that scientists and engineers can use in research with the accompanying ethical issue (italicized):

1. How is the laboratory notebook structured and what provisions are there to have the entries signed and dated by a witness?
   Ethical issue: should be signed and dated by a witness.
2. Did the laboratory notebook include changes in the views of the researcher relating to the

subject being researched, data, theory, and the method?
Ethical issue: Omitting to include such changes.
3. Are there notations relating to new ideas from the literature?
Ethical issue: Failure to acknowledge the sources.
4. What are the controls over having sufficient information?
Ethical issue: Failure to acknowledge the need for further information.
5. Which methods or combination of methods were used to collect data and apply to date workup?
Ethical issue: Using methods that will gave results that are in keeping with, and support, the theory of the researcher.
6. What are the data requirements for the research?
Ethical issue: Acquiring data from other researchers even if it means an invasion of their work (without permission) and without acknowledgement of the source.
7. What are the limitations of the research?
Ethical issue: Deliberately claiming fewer limitations once the theory has been seemingly proven.
8. How is the research problem defined?
Ethical issue: The issue of using a definition that fits the preliminary data rather than the original project definition.
9. Has the available literature been reviewed extensively and carefully for prior work?
Ethical issue: Selective reviewing for preferential papers and omission of other papers that may point the way for further work or refute the researcher's theory.

10. Is your study original in terms of methods, equipment, data generation, and procedures?
    Ethical issue: Claiming originality without a clear basis or failure to acknowledge prior work.
11. Which sampling techniques were used?
    Ethical issue: Deliberately excluding standard methods of sampling because data acquired by these methods may point unfavorable to the researcher's theory.
12. Should a research proposal include all or one of the following: title, abstract, background, all of the relevant literature, data collection methods, and implications?
    Ethical issue: failure to disclose all of the relevant information in the proposal and knowing the outcome of the research because of undisclosed data.
13. Is there a plan analyzing and interpreting the data?
    Ethical issue: Deliberately omitting an unbiased plan and/or omitting some of the data that do not support the theory.
14. Is there a plan to repeat experiments or field work if more data are required?
    Ethical issue: Fabricating additional data when the researchers should go back into the laboratory to repeat experiments, to do additional experiments, or perform more field work.
15. Is the researcher willing to seek evidence that might dispute his theory?
    Ethical issue: Failure to seek alternate evidence and/or ignore evidence that contradict the researcher's theory.
16. Is the researcher willing seek pursue an alternate theory on the basis of deviant data (e.g., flyers on an x-y plot of the data)?

Ethical issue: Ignoring or deleting any such information that threatens the theory.

17. By what means will the researcher would you analyze his data?
Ethical issue: Deliberately analyzing data in a manner which supports the theory and/or data which cannot be replicated.
18. Does the researcher show originality in techniques and procedures used to conduct the study, in exploring the unknown, in using the data, and in outcomes of the study?
Ethical issue: Claiming originality for the work without a rationale for making such claims.
19. Dues the data in the laboratory notebook help to align thinking and provide ideas for future study?
Ethical issue: Make inferences/judgments without serious consideration of the true nature of the data.
20. Is the researcher sure the he is not simply empathizing (agreeing) with the work of others under whose supervision he works and mirroring the supervisor's experience and data?
Ethical issue: Losing rational and emotional balance during or after the study when the data are being assessed.

References

Altman, E. 1997. "Scientific Research Misconduct." *In Research Misconduct: Issues, Implications and Strategies*. E. Altman and P. Hernon (Editors). Ablex Publishing, New York.

Barnbaum, D.R. and M. Byron, M. 2001. *Research Ethics: Text and Readings*. Prentice Hall, Upper Saddle River, New Jersey.

Bertozzi, C. (Chair). 2009. On *Being a Scientist: A Guide to Responsible Conduct in Research 3rd Edition*. National Academy of Sciences, Washington, DC.

Blader, S.L. and Bobocol, D.R. 2005. "Wanting is Believing: Understanding Psychological Processes in Organizational Justice by Examining Perspectives of Fairness." *In What Motivates Fairness in Organizations?* S.W. Gilliland (Editor). Information Age Publishing, Charlotte, North Carolina.

Bok, D. 2006. *Universities in the Marketplace*. University of Princeton, Princeton, New Jersey.

Branscomb, L.M. 1985. Integrity in Science. American Scientist, 73(Sept/Oct): 421.

Cassell, J. 1982. "Does Risk-Benefit Analysis Apply to Moral Evaluation of Social Research?" *In Ethical Issues in Social Science Research*. Johns Hopkins University Press, Baltimore, Maryland.

Fleddermann, C.B. 2008. *Engineering Ethics 3rd Edition*. Pearson Prentice Hall, Upper Saddle River, New Jersey.

Gibilisco, S. 2004. *Statistics Demystified*, McGraw-Hill, New York.

Gilliland, S.W. and Paddock, W.L. 2005. "Images of Justice: Development of Justice Integration Theory." *In What Motivates Fairness in Organizations*? S.W. Gilliland (Editor). Information Age Publishing, Charlotte, North Carolina.

Hamilton, Neil. (2002). *Academic Ethics*. Praeger Press, Greenwood Publishing Group, Santa Barbara, California.

Heilprin, J. 2003. "Truth in Science – Goal for New Interior Code." *Casper Star-Tribune*, May 31, page A7.

Hernon, P., and Calvert, P.J. 1997. "Research Misconduct as Viewed from Multiple Perspectives." *In Research Misconduct: Issues, Implications and Strategies*. E. Altman and P. Hernon (Editors). Ablex Publishing Corporation, New York.

Hileman, B. 2005. "Research Misconduct." *Chemical & Engineering News*. American Chemical Society, Washington, DC. 83(35): 24–26.

Huff, D. 1954. How to Lie with Statistics. W.W. Norton and Company Inc., London, United Kingdom.

Kearney, R. 1999. "The Crisis of the Image: Levinas' Ethical Response." *In The Ethics of Postmodernity: Current Trends in Continental Thought edited by G.B. Madison and M. Fairbean*. Northwestern University Press, Evanston, Illinois.

Kerr, C. 2001. The *Uses of the University*. Harvard University Press, Cambridge, Massachusetts.

LaFollette, M.C. *Stealing into Print: Fraud, Plagiarism, and Misconduct in Scientific Publishing*. University of California Press, Berkeley, California.

Mark, M.M. and Gamble, C. 2009. "Experiments, Quasi-Experiments and Ethics." *In Handbook of Social Research Ethics*. D.M. Mertens and P.E. Ginsberg (Editors). Sage Publications, Thousand Oaks, California.

Martin, M.W., and Schinzinger, R. 2005. *Ethics in Engineering 4th Edition*. McGraw Hill, New York.

McDowell, B. 2000. *Ethics and Excuses: The Crisis in Professional Responsibility*. Quorum Books, Westport, Connecticut.

Mertens, D.M. and Ginsberg, P.E. 2009. "Ethical Issues in Research Practice." *In Handbook of Social Research Ethics*. D.M. Mertens and P.E. Ginsberg (Editors). Sage Publications, Thousand Oaks, California.

NAS. 1992. *Responsible Science: Ensuring the Integrity of the Research Process*: National Academy of Sciences, National Academy Press, Washington, DC.

Neave, G. 1988. "On Being Economical with University Autonomy: Being an Account of the Prospective Joys of a Written Constitution." *In Academic Freedom and Responsibility*. M. Tight (Editor). Open University Press, Milton Keynes, Buckinghamshire, England.

NRCNA. 2002. *Integrity in Scientific Research: creating an environment that promotes responsible conduct*. National Research Council of the National Academies. The National Academies Press, Washington, DC.

Ogden, T. 1999. *Control Magazine*. Putnam Publishing Co., New York. March, page 41.

Pocklington, T., and Topper, A. 2002. No Place to Learn: Why Universities Aren't Working. UBC Press, Toronto, Ontario, Canada.

Reagan, C.E. 1971. *Ethics for Scientific Researchers, 2nd Edition*. Thomas Publishers, Springfield, Illinois.

Resnik, D.B. 2000. "Statistics, Ethics, and Research: An agenda for education and reform." *Accountability in Research*, 8: 163–188.

Roy, R. 1999. "Comment. Chemistry in Britain." *Royal Society of Chemistry*, London, United Kingdom.

Ryan, K.J. (Chair) 1995. "Integrity and Misconduct in Research." *Commission on Research Integrity*. United States Department of Health and Human Services, Washington, DC.

Ryan, J.F. 2002. "Fraud." *Today's Chemist at Work*. American Chemical Society, Washington, DC. Volume 11(11): 9.

Shrader-Frechette, K. 1994. *Ethics of Scientific Research*. Rowman and Littlefield Publishers Inc., New York.

Smith, R.B. 2008. *Cumulative Social Inquiry: Transforming Novelty into Innovation*. The Guilford Press, New York.

Speight, J.G. 2002. *Handbook of Petroleum Product Analysis*. John Wiley & Sons Inc., Hoboken, New Jersey.

Tight, M. 1988. "Editorial Introduction." *In Academic Freedom and Responsibility*. M. Tight (Editor). Open University Press, Milton Keynes, Buckinghamshire, England.

Werner, B. 1995. *Judgments of Responsibility: A Foundation for A Theory of Social Conduct*. The Guilford Press, New York.

Whitbeck, C. 1998. *Ethics in Engineering Practice and Research.* Cambridge University Press, New York.

Wolf, A., Turner, D., and Toms, K. 2009. "Ethical Perspectives in Program Evaluation." *In The Handbook of Social Research Ethics.* D.M. Mertens and P.E. Ginsberg (Editors). Sage Publications, Thousand Oaks, California.

8

Publication and Communication

8.1 Introduction

Publication plays a critical role in the advancement of science and engineering by communicating knowledge from the researcher(s) to the larger scientific community (Davis, 1997). One might say that science and engineering communication does not truly exist until the data are published, at which time the publication becomes a public commodity. The exchange of information through publication is an essential part of doing science and engineering, a public good, and, for some, a moral imperative. It is important, then, that scientific societies, as major publishers of science and engineering, take initiatives to preserve the integrity of the process that certifies and communicates research (Beckett, 2003).

Publication of papers in peer-reviewed journals is the predominant form of publication for scientists and engineers, and usually such journals have the highest readership.

However, journals vary enormously in their prestige and importance, and the value of a published article often depends on the journal.

Peer review is a general term that is used to describe a process of self-regulation by scientists and engineers (as well as for many other professions) as a means of evaluation of paper before publication; this involves review qualified individuals in the relevant scientific or engineering field. Peer review methods are employed to maintain standards, improve performance, to verify whether the work satisfies the specifications for review, to identify any deviations from the standards, and to provide suggestions for improvements.

In general, the professionals who are biased towards theory tend to produce data that are often abstract and the intellectual contribution is expressed in the form of theories with proof. As a result, publication on the proceedings of a conference may be the only outlet for their efforts after which publication in a *reputable* journal may be possible but only with considerable efforts or, for various reasons, may not be possible at all. For the non-academic scientist and engineer, there is the medium of publication of the material as a *company report*. This can be a worthwhile method for circulating one's work throughout the company. But, the importance of the work to the young scientist and engineer can, again, be diminished while the names of a supervisor, and any other persons higher up the food chain, are included as co-authors.

Publication of data in the proceedings from a conference often results in a shorter time to print. This follows from the opportunity to describe completed or partly completed work before peer scientistsor engineers receive a more complete review than the type of review that is typical for a journal. At a conference, the audience asks general and specific questions to the presenter that often provides recommendations for further work or a new line of investigation. Overall, this will help the presenter to finalize the document for publication in the proceedings (where the proceedings are published post-conference), using any lame excuse that comes to mind. This, surely is a breach of ethics, which may involve untruths or merely laziness.

In addition, the young Assistant Professor also has to acquire research funding and may even have to pass his reports/papers through a review committee prior to publication. This review committee will be made up of senior members of staff who, for many reasons that are often difficult to follow, can give the young professor a flowing performance report or a report that is somewhat less than glowing. It is at this time, if the latter is the case, that the young professor can feel that he is suffering rejection by one's colleagues.

The educated young professional scientist and engineer wonders if he is merely a pair of hands (for a overbearing supervisor, an overbearing department head or jealous colleagues) and not supposed to be given credit for the ability to think and solve a problem. Performance suffers and, with repeated negativism towards publication, the young professional starts to lose interest in the organization.

Lack of recognition for hard and intelligent work is a killer and getting the best out of any such scientists and engineer becomes an impossible dream.

There is an extremely important role for the scientific societies in developing authorship policies for their members. The societies must also make sure that their members know of the existence of their policies and how to interpret them. Regular continuing educational efforts are imperative. There is also the possibility that scientific societies could work together to establish a uniform policy that would hold across disciplines. This would be advantageous to those engaged in interdisciplinary research collaborations.

While the general definition of scientific misconduct includes fabrication, falsification, and plagiarism, the scientific community is charged with considering standards for other practices. In publication practices, that encompasses such matters as authorship credit, duplicate publication, accurate representations of the data presented, and peer review.

Generally, the following criteria need to be observed when compiling data for publication:

1. all persons designated as authors should legally qualify for authorship, and
2. each author should have participated sufficiently in the work to take public responsibility for the content.

Furthermore, authorship credit should be based only on substantial contributions to:

1. either the conception and design or the analysis and interpretation of data;
2. drafting the article or revising it critically for important intellectual content; and
3. final approval of the version to be published.

Conditions 1, 2, and 3 must all be met. Other contributors should be listed in an appendix or footnote. Editors may ask authors to describe their contribution(s).

However, publishing is undergoing redefinition as electronic publications and there are both opportunities and pitfalls associated with electronic publishing. The immediacy, impermanence and global reach of electronic publishing mean that new, expanded audiences can be reached. In addition, digital technology may make it easier to misrepresent data or alter graphic representations. Societies could make a valuable contribution by encouraging cross-disciplinary discussion of these matters among researchers and those involved in publishing. Guidelines for responsible conduct in the electronic communication and electronic publication of scientific research must be developed and implemented, and the societies can play a pivotal role in their promulgation and implementation.

Generally, publications are variable but, in general, constitute:

1. They constitute a report and a record of the activity of researchers.
2. They constitute the references research builds upon.
3. They constitute the data governments, organizations, or society can refer to when facing issues that have a major scientific component.
4. They constitute a way of evaluating the scientific and engineering activities of individual researchers.

The quality of publications, in terms of scientific integrity, is therefore essential for research to be conducted in an efficient and responsible way and for a transparent communication between researchers and society.

Science and engineering disciplines are flourishing because of communication, a much broader concept than publishing. Hopefully, before the article will be published, researchers will have had extensive discussions with their peers to share their views, ideas and opinions in order to check the validity of their claims. Unfortunately, in recent times the advancement of knowledge has not appeared to be a top priority for many scientists and engineers. Fame and fortune have become the focus of the researchers and publication of data that have not been confirmed or data that have been made up have seen the light of day in scientific and engineering journals.

Many opportunities and concerns are at play in scholarly publication and communication. These result from capabilities afforded by new technologies, pressures associated with the publish-or-die message that is forced on many scientist and engineers in academia or the invent-or-die message that is forced on many industrial scientist and engineers.

While the unethical behavior of scientists and engineers cannot be blamed on, the publish or die message, or on the, invent or die message, the pressure placed on the shoulders of many individuals by either of these messages may be a contributing factor. Not that anyone found guilty of unethical behavior should be excused because of such a message but it may be time to change the message – if that is at all possible.

The young assistant professor, who is excused from being reprimanded for unethical behavior because he is a young professor seeking funding for a research project, is also not a valid excuse for lack of disciplinary action. In fact, one might ask if those exalted academics promoting and accepting such an excuse are not also guilty of unethical behavior because they have condoned the professor's actions.

In fact, the lack of willingness of the (academic or industrial) faculty to change is a key barrier to reducing and perhaps eliminating unethical behavior in science and engineering and engineering.

While there are claims that gross scientific misconduct are assumed to be rare (however, please see Chapter 9), subtler forms of unethical behavior are becoming more common (Ritter, 2001). Misappropriated credit in publications, for example, can lead to some of the most contentious conflicts in the academic world. Currently, in academia, publication of research data has become more competitive because universities and organizations are more focused on intellectual property and rights of ownership. In addition, research that is sponsored by commercial entities is usually controlled; it is the commercial organization which determines whether and how results are published is no longer an academic issue.

Guidelines for responsible conduct in the communication and publication of scientific research must be developed and implemented, and the societies can play a pivotal role in their promulgation and implementation.

8.2 The Scientific and Engineering Literature

The *scientific and engineering literature* comprises scientific and engineering publications (journals) that report original empirical and theoretical work in scientific and engineering disciplines. University researchers favor publication in such journals while their industrial counterparts may have to focus on patents.

Currently, peer-reviewed journal articles remain the predominant publication type, and have the highest prestige. However, journals vary enormously in their prestige and importance, and the value of a published article depends on the journal. The status of conference proceedings depends on the discipline; they are typically more important in the applied science and engineering, especially for industrial scientist and engineers.

In many scientific and engineering disciplines, advancement depends upon publishing in so-called "high-impact" journals, most of which are English-language journals. Scientists and engineers with poor English writing skills are at a disadvantage when trying to publish in these journals, regardless of the quality of the scientific study itself. Yet many international universities require publication in these high-impact journals by both their students and faculty. One way that some international authors are beginning to overcome this problem is by working with technical copy editors who are native speakers of English and specialize in editing texts written by authors whose native language is not English to improve the written quality to a level that high-impact journals will accept.

This is necessary because a scientific article has a standardized structure, which varies only slightly in different subjects. Ultimately, it is not the format that is important, but what lies behind it, the content, and how well the content is explained by the authors. In most cases, several formatting requirements need to be met. For example, the title should be concise and indicate the contents of the article. Most important, the names and affiliation of all authors are given – because of case of misconduct, the publisher may require that all co-authors know and agree on the content of the article.

The format of the paper is also subject to certain requirements there should be an *abstract* (a one-paragraph summary of the work, usually less than a specified number of words) which is intended to serve as a guide for determining if the article is pertinent to potential readers. Following the *abstract*, there is an *introduction* in which previous works relevant to the work in the paper should be presented in the context of previous scientific or engineering investigations, by citation of relevant documents in the existing literature. Then follows the *experimental* section in which the method and materials are described after which the data are presented in the *results* section. Interpretation of the meaning of the results is usually addressed in a *discussion* section and the conclusions should be based on previous literature and/or new empirical results, in such a way that any reader with knowledge of the field can follow the argument and confirm that the conclusions are sound. The final section is *references* (literature cited) section in which the sources cited by the authors are listed in the format required by the journal.

8.3 The Journals

Science and engineering is supposed to be a project centered on building a body of reliable knowledge about the universe and how various pieces of it work. This means that the researchers contributing to this body of knowledge for example, by submitting manuscripts to peer reviewed scientific journals, are supposed to be honest and accurate in what they report. They are not supposed to make up their data, or adjust it to fit the conclusion they were hoping the data would support. Without this commitment, science and engineering turns into creative writing with more graphs and less character development.

Because the goal is supposed to be a body of reliable knowledge upon which the whole scientific community can draw to build more knowledge, it's especially problematic when particular pieces of the scientific literature turn out to be dishonest or misleading. Fabrication, falsification, and plagiarism are varieties of dishonesty that members of the scientific community look upon as high crimes. Indeed, they are activities that are defined as scientific misconduct and (at least in theory) prosecuted vigorously.

It is to be hoped that one consequence, of identifying scientists and engineers who have made dishonest contributions to the scientific literature, would be removal of dishonest contributions from the literature. Yet, whether that hope is realized is an empirical question.

Journals occasionally report on notorious research integrity violations, summarizing information from scientific misconduct investigations, and noting the affected publications. Many other lesser-known cases of fraudulent publications have been identified in official reports of scientific misconduct; yet there is only a small body of research on the nature and scope of the problem, and on the continued use of published articles affected by such misconduct.

The standard for being caught is having an official finding of misconduct against the authors or perpetrators of the misconduct. In part, this is because such a finding usually includes consequences connected to publications that may embody the dishonesty toward fellow scientists and engineers.

Not every retraction is the result of a finding at the end of an inquiry into misconduct. Conversely, in situations where there has been an inquiry into misconduct and the finding is that there has been misconduct that requires correction of the literature (via a correction or a retraction), it is hoped that the coauthors of the paper would be subject to the appropriate action. However this is not always the case. Numbers are not always available, but many perpetrators of misconduct are exonerated with time (time is a great healer and memory scrubber) and continue to practice science and engineering as if nothing had ever happened.

It is also to be hoped that scientific journals would recognize their interest in serving their readers by ensuring the scientific quality of the articles they publish. The prepublication screening (via peer review and editorial oversight) can do part of the job, but even in situations where there is nothing like misconduct on the part of authors, occasionally honest mistakes are discovered after publication. Some journals even have a policy that would prevent authors who become aware of such mistakes from communicating the relevant information to their fellow scientists and engineers who have access to the published work now known to be mistaken, whether through the publication of a correction or a retraction.

In any case, the present study points to policies and facts on the ground that might make us worry about how completely errors in the scientific literature (whether honest mistakes or intentional deceptions) are corrected.

Publishing may not always move quickly, but surely that three years is sufficient to communicate to the scientific community that draws on the literature whether a particular piece of that literature is not as reliable as it was first thought to be.

For example, the published findings of misconduct in the *NIH Guide for Grants and Contracts* and in the ORI Annual Reports for 1991–2001, one hundred and two articles were identified as needing retraction or correction. There were forty one researchers whose misconduct was tied to the one hundred and two articles, nineteen of them identified as responsible for a single problematic paper and twenty two were responsible for two or more problematic papers. One of those forty one researchers was responsible for a ten articles that were in need of retraction or correction.

Furthermore, of those one hundred and two articles, seventy nine reported results that were fabricated, falsified, or misrepresented; two contained plagiarism; sixteen gave inaccurate reports of the methodology the researchers actually used; and five reported "results" from fabricated experimental subjects. Just over half of the forty one researchers (responsible for fifty three of the flagged articles) accepted the findings of misconduct, while five were recorded as disagreeing with the findings or denying responsibility for the misconduct. The other misconduct findings didn't record the response of the respondents to the findings. By the time the findings of misconduct were published, corrigenda (corrections) had already been published for thirty two of the flagged articles and sixteen more were in press. Retractions or corrigenda needed to be published for another forty seven of the flagged articles.

This leaves seven of the articles flagged (as reporting results that were fabricated, falsified, or misrepresented, or as containing plagiarism, or as giving inaccurate reports of the methodology the researchers actually used, or as reporting results from fabricated experimental subjects) for which the administrative actions did not specifically call for correction or retraction. However, it's not unreasonable to think that articles flawed in these ways ought to be corrected or retracted; in order to protect the reliability of the scientific literature and the trust scientists and engineers need to be able to place in the reports published by their fellow scientists and engineers.

Potentially, this is a problem.

The thought goes to the means by which the continued citation of research affected by scientific misconduct can be reduced. More prominent labeling in the literature is desirable to alert users to notices of retraction and errata. This could take the form of larger or bold fonts for these notices. Alternatively, or in addition, a prominent placement of the word *retraction* on the first page of such articles would be useful, because once a user downloads an article, any notices related to retraction of the article may be are left behind.

Some of the problem, in other words, may be due to the vigilance (or lack thereof) displayed by those using the scientific literature, but some of it may come down to the extent to which that scientific literature is accessible to the researchers.

Laboratory directors and principal investigators often do not check every detail of the work by students or their junior colleagues but the onus is on the director or principal investigator work and trust that the work is accurate (Ritter, 2001). Thence, it falls upon the shoulders of universities, journal editors, and reviewers to determine that the directors or principal investigator has submitted a manuscript that is accurate and true.

Weeding out such problematic papers out of the pool of scientific literature that researchers cite may require journal editors, manuscript authors, and even journal readers to take on more responsibility. For example, before authors submit a manuscript for publication (either initially or after the last set of revisions), they must ensure that none of the sources they cite have been retracted or corrected. Failing to exercise such vigilance could inadvertently render the paper a problematic, especially if the paper depends in part on another problematic paper (and so on).

However, until the scientific community is willing to recognize and practice such vigilance as a duty, it's unlikely that failing to exercise it could itself rise to the level of scientific misconduct.

8.4 Data Manipulation for Publication

Data manipulation for publication must not be used as an excuse for unethical behavior; it is recognized the Internet has changed the way science and engineering is done, particularly when it comes to publication. Manuscripts are now submitted, reviewed and authors notified electronically. But although the efficiency and speed of the peer-review process has increased, a set of attendant issues has arisen (Oxender Burgess, 2004).

Specifically, it is now easier to detect breaches of ethical behavior than ever before. As evidence, the number of reported ethical problems involving publications of the American Physiological Society (in 14 separate journals) rocketed from an average of less than one a year before 1999 to more than 50 a year in 2004, when all of the society's publications became available online.

When evaluating a manuscript, a reviewer no longer has to trek several blocks to the library to scour the printed journals in search of a paragraph or a figure that seemed familiar. All that has to be done now is to type a few keywords into an appropriate search engine and, hey presto, all the relevant articles will appear on your desktop. Although for the more meticulous, anti-plagiarism software is available for free download.

Other, more draconian misconduct-detection measures are aimed at identifying image manipulation. These are currently being considered and even implemented by some journals.

It is all too easy for authors to manipulate images for publication. For example, digital image-processing programs make it a simple matter to remove *non-specific* bands can be easily removed from the final figure. Since the foundation of good science and engineering is accurate, reliable and reproducible data, then images that are less than perfect must be accepted. If images are manipulated to enhance presentation, the integrity of the scientific and engineering enterprise may be compromised and erase the trust that the public places in our work.

Two of the most obvious ways that data distortion can be produced are by altering the relationship between the horizontal and vertical scales and the use of different scales on the left and right hand sides of the graph or along the bottom of the graph.

Quality data, reported in a clear and concise format, may be the dominant factor in determining whether the information presented is used or disregarded as worthless, or worse, intentionally misleading. Only by creating and maintaining some sort of organizational ethical guidelines and educating data graphics designers on the effect poorly designed graphics have on people's perceptions can we ensure quality data presentations.

However, it may not be prudent for journal editors to seek out potential misconduct in every submitted manuscript. That would impose an unnecessary confrontational relationship on authors and publishers, even before the process of peer review began. However, journal editors may be responsible for ensuring the integrity of the scientific record but if scientists and engineers do not police their own actions and actively instruct students in proper behavior, someone else will and the consequences may be unpalatable.

8.5 Detecting Falsified Data

With the popularity of the Internet as an information source, there are now several tools available to aid in the detection of plagiarism and multiple publications of the same paper within the scientific and engineering literature.

In addition to the various software packages, other tools which may be used to detect falsified data include error analysis. Error analysis is based on the principle that experimental measurements generally have a small amount of error, and repeated measurements of the same item will generally result in slight differences in readings. These differences can be analyzed, and follow certain known mathematical and statistical properties. Should a set of data appear to be too faithful to the hypothesis, i.e., the amount of error that would normally be in such measurements is not evident, a conclusion is that the data may have been subject to manipulation (forged). Error analysis alone is typically not sufficient to prove that data has been falsified, but it may provide the supporting evidence necessary to confirm suspicions of misconduct.

8.6 Peer Reviewers and Their Duties

Peer review is the means by which scientific and technical manuscripts submitted for publication in journals undergo quality control in the form of a check on technical quality, the lack of flaws in the data, and the validity of the conclusions drawn from the data. The lack of peer review is what makes most technical reports and World Wide Web publications unacceptable as contributions to the literature. The relatively weak peer review often applied to books and chapters in edited books means that their status is doubtful, unless an author's personal standing is so high that his or her prior career provides an effective guarantee of quality. Even then, it is not beyond a reputable scientist or engineer writing a book that is of very low quality both in terms of the writing and the content.

Formal peer review is in flux and likely to change fundamentally owing to the emergence of institutional digital repositories where scholars can post their work as it is submitted to a print-based journal. Though this does not prevent peer review, it permits an unreviewed copy into general circulation.

The peer review process is central to scholars' perception of quality in a journal that its retention is essentially a *sine qua non* for any method of archival publication, new or old, to be effective and valued (Harley et al., 2007). Peer review is *the* hallmark of quality that results from external and independent valuation. It also functions as an effective means of winnowing the papers that a researcher needs to examine in the course of his or her research.

Peer review is an essential factor when faculty were asked about their perceptions of both standard and newer forms of publication, disadvantages of newer forms of publication, where to publish to make a name for oneself in the field, and, of course, when we asked about peer review specifically.

There is a strong tendency for many members of the research community to equate electronic-only publication with lack of peer review, despite the fact that there are many examples to the contrary. Because of the very nature of peer review, this factor inhibits publication in electronic-only venues even among those who are aware of the existence of fully peer-reviewed e-journals. Simply put, they know that the individuals reviewing their work for advancement may well not have that awareness.

It will be important to try to separate the issue of peer review for newer, electronic journals from those issues associated with the fact that most such journals are simply new and not yet well established. To some degree, however, peer review and the means of publication and dissemination can be separated. For example, there are authors whose work is peer reviewed and published in prestigious print journals, but who also retain rights to place the article on their own Web site. The result is that the work appears to be accessed far more often on the Web site than in the published print journal.

There is a growing tendency to rely on secondary measures associated with peer review. These include perceived journal quality, selectivity, and/or stature; whether papers or keynote lectures for conferences are invited; and the growing reliance on editors of university presses and reviewers for journals to evaluate scholarly work. Even though reviewers for university presses are academic faculty, the editor exerts much more independent judgment than is typical for peer-reviewed journals published by scholarly societies. In some cases, the impact factor may also serve as a gauge of quality, a development that many view as problematic, as long as the impact factor is not a measure invented by one publisher for application to journals published by that particular company.

For example, what does the impact factor really mean if papers from journals of Publisher A have higher impact factors when judged using the system that originated from Publisher A? This question is skirting the issue of unethical behavior; but isn't such a system setup to make publishers A's journals more in demand than the journals of other publishers?

The peer-review process is more complicated for compound disciplines which cross over between science and engineering because many such fields are relatively nascent, and therefore, result in small, specialized communities of scholarship.

Scientists and engineers, in these interdisciplinary fields, often prefer to publish within a single traditional discipline because the most highly respected and recognizable outlets reside there; however, divergent expectations (ranging from quantity to methodology to writing style) and standards (especially with regard to quality) among fields often make it difficult for reviewers in standard fields to judge submissions from compound disciplines. Interdisciplinary publications may address this concern more readily as they become more prestigious. However, in such fields, the utilization and perception of peer review is particularly complicated, given the prominence of student-edited law reviews.

8.7 Duties and Responsibilities of a Journal Editor

Descriptions for duties and responsibilities of an editor of a scientific or engineering journal are scarce.Generally, the editor of a peer-reviewed journal is responsible for deciding which of the articles submitted to the journal should be published, often working in conjunction with the relevant society (for society-owned or sponsored journals). The validation of the work in question and its importance to researchers and readers must always drive such decisions. The editor may be guided by the policies of the journal's editorial board and constrained by such legal requirements as shall then be in force regarding libel, copyright infringement and plagiarism. The editor may confer with other editors or reviewers (or society officers) in making this decision.

More specifically, the Journal Editor (or Editor in Chief, as the title may indicate) has final decision-making authority on, and is responsible for the appropriate delegation of Editorial Board responsibilities related to, the scientific, engineering, and other editorial content of the journal: including solicitation and acceptance or rejection of manuscripts; selection of editorial board members and reviewers; and the approach to correspondence with authors, reviewers, and readers.

In order to accept these duties and responsibilities, the editor must be skilled in the areas of scholarship covered by the journal. In short, the editor of a scientific or engineering journal has the added responsibility to check the technical soundness and technical quality of the content. For this, the editor is required to have the technical skills and up to date knowledge of the area of scholarship covered by the subject matter of the journal.

For example, the scientist or engineer who takes work from a previously unknown paper (say one published in a foreign journal that is little-known and little-read in the west) should give credit to the original work and not ignore it on the basis of personal likes and dislikes, or even on the basis of the personal likes and dislikes of his supervisor/mentor.

An editor who is up to date in his own area of scholarship would immediately recognize such omissions.

Thus, the assumption (often correct, but not always) is that the editor chosen for a journal is the most appropriate scientist or engineer because the editor is the "gatekeeper" or "watchdog" for the journal; anything published in the journal must pass across the editor's desk and *must* be reviewed by the editor. Authors who submit manuscripts to the journal for possible publication are often swayed by the qualification and knowledge of the editor – the reputation of the editor is a direct influence on the reputation of the journal.

The editor is, in fact, the quality control officer for the journal where a check is made on the content (authenticity and relevancy to the topic), language (grammar and content flow) and aesthetics (photos, images, sound, audio and video) of the articles or documents appearing on the specified medium. An editor is required, with the consent of the relevant authors, to change, modify, paraphrase or condense the content in order to enhance its quality and approve or reject the piece based on preset grounds. The job of the editor also involves relationship building and communication with the author. The editor is required to use his creative skills and human resource skills to maintain a cordial relationships with authors whose article are rejected.

Above all, journal editorship is not a resumé builder! In fact, the success of a journal will depend on the performance of the editor and it is the job of the editor to shepherd the journal through lean times and through good times (McHugh, 1998). For those who seek to build a resume by including journal editorship as a glowing one-liner, it is recommended that they seek early retirement.

In addition being the good shepherd, the other major responsibility of the editor is the administration of the peer review process.

The peer review process assists the editor in making editorial decisions and through the editorial communications with the author may also assist the author in improving the paper. The process is an essential component of formal scholarly communication, and lies at the heart of the scientific method. All scientist and engineers who wish to contribute to publications should be willing (perhaps even have an obligation) to act as peer reviewers.

Any selected reviewer who feels unqualified to review the research reported in a manuscript or knows that its prompt review will be impossible should notify the editor and excuse himself from the review process.

In addition, the journal peer review process has three purposes. The first is quality control, to eliminate major errors in papers and unsuitable papers. Second, the review process should ensure fair treatment of all authors (not just for cronies of the editor) and especially not for paper authors or co-authors by the editor. Third, the review process encourages the publication of papers that contain new and useful findings.

Furthermore, the editor should also know the reviewers sufficiently well to know that they are qualified, impartial and fair and the confidentiality of the review process must be preserved.

Protecting intellectual property is a primary responsibility of the editor. The editor should know the reviewers well enough so that any thoughts of plagiarizing manuscripts by reviewers should not be an issue. Reviewers must not use ideas from or show another person the manuscript they have been asked to review without the explicit permission, via the journal editor, of the author of the manuscript. Advice regarding specific, limited aspects of the manuscript may be sought from colleagues with specific expertise, provided the author's identity and intellectual property remain secure.

The editor must accept it as a hard and fast rule (a rule without exception) that unpublished materials, disclosed in a submitted manuscript, must not be used in a reviewer's own research without the express written consent of the author. Privileged information or ideas obtained through peer review must be kept confidential and not used for personal advantage. Reviewers should not consider manuscripts in which they have conflicts of interest resulting from competitive, collaborative, or other relationships or connections with any of the authors, companies, or institutions connected to the submitted unpublished papers.

The editor must also maintain reviewing schedules by ensuring that reviewers meet agreed-upon reviewing deadlines. And, in order to maintain good reviewers the editor should have the means to evaluate performance of editorial review board members and coach when appropriate.

Whilst some editors consider themselves to be the all-powerful authority for publication in the journal, the editor should be willing to step down from this lofty perch and assist authors in developing articles to the fullest potential. To do this, the editor must maintain all communications with all authors and reviewers in a courteous and diplomatic manner. This also involves correspondence sent to authors in relation to checking every manuscript for completeness, references, tables, figures suitable for reproduction, legends, abstracts, permission to use copyrighted material, and mailing address for proof.

Above all, and certainly in the context of the present text, the editor must not be involved in positions where conflicts of interest can arise so that all decisions are beyond reproach.

A conflict of interest may exist when a manuscript under review puts forth a position contrary to the reviewer's published work or when a manuscript author or reviewer has a substantial direct (or indirect) financial interest in the subject matter of the manuscript. A conflict of interest may also exist when a reviewer knows the author of a manuscript. The editor should ensure that such conflicts do not occur and that he is also beyond reproach.

The editor should also assure that proper acknowledgment of the work of others must always be given. Authors should cite publications that have been influential in determining the nature of the reported work. Information obtained privately, as in conversation, correspondence, or discussion with third parties, must not be used or reported without explicit, written permission from the source. Information obtained in the course of confidential services, such as refereeing manuscripts or grant applications, must not be used without the explicit written permission of the author of the work involved in these services.

The editor should recognize that a conflict of interest does not exist when an author disagrees: with a reviewer's assessment, that a problem is unimportant, or disagrees with an editorial outcome.

When an author discovers a significant error or inaccuracy in his own published work, it is the author's obligation to promptly notify the journal editor or publisher and cooperate with the editor to retract or correct the paper. If the editor or the publisher learns from a third party that a published work contains a significant error, it is the obligation of the author to promptly retract or correct the paper or provide evidence to the editor of the correctness of the original paper.

An editor should take reasonably responsive measures when ethical complaints have been presented concerning a submitted manuscript or published paper, in conjunction with the publisher. Such measures will generally include contacting the author of the manuscript or paper and giving due consideration of the respective complaint or claims made. Actions taken by the editor may also include further communications to the relevant institutions and research bodies; and, if the complaint is upheld, the publication of a correction, retraction, expression of concern, or other note, is required. Every reported act of unethical publishing behavior must be investigated, even if it is discovered years after publication.

References

Beckett, R. 2003. "Communication Ethics: Principle and Practice." *Journal of Communication Management*, Henry Stewart Conferences and Publications Ltd., London, 8(1): 43–44.

Davis, M. 1997. *Scientific Papers and Presentations*. Academic Press, New York.

Harley, D., Earl-Novell, S., Arter, Lawrence, S., and King, C.J. 2007. "The Influence of Academic Values on Scholarly Publication and Communication Practices." *Journal of Electronic Publishing*, Volume 10, Issue No. 2.

McHugh, J.B. 1998. "Responsibilities of a Journal Editor." *McHugh Publishing Reprint* 22. http://www.johnbmchugh.com/pdfs/P-2%20 22%20Responsibilities%20of%20a%20Journal%20Editor.pdf

Oxender Burgess, D. 2002. "Graphical Sleight of Hand." *Journal of Accountancy*, American Institute of Certified Public Accountants, New York, NY, February. 193(2): 45–47.

Ritter, S.K. 2001. "Publication Ethics: Rights and Wrongs." *C&E News*, American Chemical Society, Washington, DC. November 12, page 24–31.

9

Enforcement of Codes of Ethics

9.1 Introduction

The occurrence of misconduct (cheating) starts early in the life of a student, usually at school, then following through to university, and into adult life (Chapter 1 and Chapter 4) (Marcoux, 2002; Carpenter et al., 2004). Whatever the reasons or rationale given for cheating or plagiarism, and there are many (Relman, 1989; Woodward and Goodstein, 1996; Brown, 2007) – they should not be accepted. It is not sufficient to give the miscreant to merely accept an apology from a knowing cheater followed by a slap on the wrist. This type of action only serves to propagate the problem.

The news media are often criticized for reporting cheating and misconduct in noticeable headlines (Chang, 2002). But if they do not which societies or organizations (i.e., universities) will stand up for what is right and report cheating (misconduct) to its fullest extent? They will make a stance, if forced into it by the news media!

It is essential that any organization enforces the code of ethics that has been adopted by the organization. If this is not done, anarchy in the form of misconduct and any other forms of unethical behavior that can be conjured up will be prevalent (Frankel, 1989; Schwartz, 2001; Fleddermann, 2008).

In fact, the authority of any society or organization to discipline the membership should be clearly stated in the Code of Ethics. Each society should have the ability to determine the Rules of its Proceedings, punish its members for disorderly behavior, and, with the concurrence of (usually) a specified majority, expel a member. Through the years, disorderly behavior has become synonymous with improper conduct such as disloyalty, corruption, and financial wrongdoing, particularly for personal gain.

Furthermore, Section 8B2.1.(b)(5)(B) of the United States Sentencing Commission's Federal Sentencing Guidelines states, "the organization shall take reasonable steps – to evaluate periodically the effectiveness of the organization's compliance and ethics program." Should a company or organization be involved in a criminal proceeding, the extent to which the company (or organization) takes steps to ensure that their compliance and ethics program is effective should be considered during the sentencing phase. The same should apply to any political or commercial organization.

Not measuring the effectiveness of a program only increases the risk and exposure for the company. In dealing with an organization, customers, suppliers, employees and communities would be wise to ask questions about the organization's intent and commitment to integrity and proper ethical conduct.

Although there are many examples of misconduct one might ask, in the context of political and financial wrongdoing, how a politician (or anyone) might forget he has the equivalent of $10,000,000 in a foreign bank account, or how a bank manager can legally become a multi-millionaire when his salary does not allow such accumulation of wealth. Yet, such events do occur!

However, through the years, perceptions of wrongdoing or conflicts of interest by society members may have changed. What might be viewed today as blatant impropriety could have been an accepted as norm or simply ignored years ago. Thus, the Code of Ethics should be frequently updated to move with the times and maintain a cautious "over-watch" of the behavior of the members.

Finally, it is better to cut off some of the misconduct at the source by giving scientists and engineers the recognition that they deserve. Generally, scientists and engineers who believe that they deserve more recognition are more likely to falsify, plagiarize, or manipulate their data in order to report successful results. Small scale deviant practices are likely to persist because, despite the canons of scientific research scientists can always attribute small inconsistencies to unavoidable errors that accompany or infiltrate all research (Glaser, 1964; Barber, 1976). One of the major determinants of judgments of the degree of responsibility is whether a controllable act is perceived or intentionally committed or due to negligence (Werner, 1995, page 13). Since judgment can only be reliably made after some period if observation or investigation. There is a general feeling that practices have increased.

Before deciding whether an ethical crisis exists, it must be determined whether or not one of three situations exists:

1. whether ethical standards are unknown and unclear,
2. whether ethical standards are clear but ignored, or
3. whether ethical standards being followed (McDowell, 2000, page 6).

Whether or not there is a crisis in professional responsibility depends on the extent to which individuals were responsible and disciplined before acquiring professional status. The fact remains that the search for truth, knowledge and understanding of scientific and engineering phenomena pose powerful ethical demands for the individual who wants to be part of a community of individuals who call themselves scientists and/or engineers (Guba, 1990, page 145). Indeed, methodological, analytical, and ethical issues are closely interconnected (Ryan, 2009, 229); particularly so, because we have to relate with people in doing research, people whose attitudes, values, and perceptions of issues vary.

9.2 Following a Code of Ethics

A code of ethics is important in many respects but, above all, it sets the *tone* from the top of the culture of the organization (Hileman, 2005). Moreover, an effective code establishes the ethical expectations for employees and management alike, and sets forth the mechanisms for enforcement and consequences of noncompliance. When the Code is perceived as an integral component of the organization's culture, is understood, followed and enforced, it can provide protection for the organization from the actions of a *rogue employee*.

The code should set forth the process for its administration, including mechanisms to disclose and document any potential conflicts of interest or to obtain waivers from any particular policy or provision. It should also provide guidance to assist employees or the membership in evaluating specific circumstances, with the standard for behavior being: if all the facts and circumstances regarding the matter were made public, the employee or member involved and the organization should be pleased to be associated with the activity.

However, the major challenge in relation to measuring the effectiveness of a code of ethics of any organization is related to the definitions that are contained within the code.

Unless the definitions in a code of ethics are sufficiently specific, the definitions are meant to be interpreted on an as-needed basis. On the other hand, if the definitions in a code of ethics are too specific, the membership or the employees will exploit loopholes and engage in behavior that technically follows the letter of the code, but not necessarily the spirit of the code. Therefore, the language used in a code of ethics must be understandable, lacking loopholes, and promote honesty and ethical principles that are in keeping with the goals of the organization.

A code of ethics for scientists and engineers, and for any organization has been put in writing and formally adopted (Chapter 5 and Chapter 6).

Understanding a code of (professional) ethics, as a convention between professionals, we can explain why scientists engineers cannot depend on mere private conscience when choosing how to practice their profession, no matter how good that private conscience; and why scientists engineers should take into account what an organization of scientists or engineers has to say about what scientists or engineers should do (irrespective of directives from above that appear to contract honorable and ethical behavior).

The language of any code of ethics must be interpreted in light of what it is reasonable to suppose the authors of the code intended. It is to be assumed that the authors of a code of scientific or engineering ethics (whether those who originally drafted or approved it or those who now give it their support) are rational persons and had the best interests of the organization and them membership in mind when they wrote the code. Furthermore, it is reasonable to suppose that the code of ethics would not require them to risk their own reputations and that the code includes anything that would be generally considered to be immoral.

Given the above, scientists and engineers are clearly responsible for acting as the respective codes of ethics require. Scientists or engineers should behave as the code requires and should also support the code by encouraging others to do as the code requires and by calling to account those who do not.

9.3 Enforcing a Code of Ethics

An organization that wishes its code of ethics to have an impact on the actions taken can increase the likelihood of this by careful selection and presentation of the content of the code. This can also increase the likelihood of having an impact by including an enforcement provision in the code.

While code content affects actions by changing a the beliefs of a decision maker as to whether an action is ethical, an enforcement provision affects actions by making the unethical (or ethical) action less (or more) desirable to the decision maker (Lere and Gaumnitz, 2003). An organization choosing to have an enforcement provision in its code of ethics is providing additional incentives (or disincentives) to encourage (or discourage) selection of certain actions.

The enforcement provisions in a code of ethics can have an impact on the employees or membership of organization insofar as for the goal of the procedures for enforcing of any *The Code of Ethical Principles and Standards,* which is to eliminate unethical behavior, not to impose punishment. Above all, the code must be written in easy-to-read and understandable but relevant language. Two ways to improve the delivery of the message are to: increase the clarity with which the message is presented, and secondly to avoid having an excessively long code.

Furthermore, in designing an enforcement provision for a code of ethics, an organization may wish to consider how it will know if the provisions of the code have been followed and the type of mechanism that will determine if a penalty is to be levied and how the individual guilty of the unethical action will, in fact, bear the penalty that the organization has specified.

In some professions, membership in a professional organization is voluntary. In this case, membership is not associated with the ability to engage in the profession. In fact, many members of the profession may not belong to the organization. If membership in an organization is voluntary, the penalties that an organization can specify generally will not have as significant an impact on decisions made as if membership is mandatory.

The goal in improving code effectiveness through increasing *clarity* is to increase comprehension – the goal is to make sure that the position that the code framers intend to communicate is the message that the decision maker receives. This can be achieved by precisely stating each code position and the greater the clarity, the more likely it is that the decision maker will change his or her beliefs concerning the ethical nature of an action and select the action desired by the framers of the code.

The *length* of a code may have an impact on its effectiveness – having a relatively short code may increase the likelihood of changing the perceptions of the decision makers by increasing awareness of the meaning of the code. On the other hand, a long code is apt to reduce the likelihood of the decision maker being aware of a specific position in the code. Providing each individual within the company or organization with a code of ethics that is many pages long is likely to result in few if any of the individuals reading the code. Even if individuals have read the code, long codes reduce the likelihood that they will retain enough of the guidance provided in the code to have a major impact on decisions made.

Enforcement of a code of ethics is more likely to be straight forward when the code is clear and understandable and leaves no room for misinterpretation or interpretation that will benefit the person accused of misconduct. The code of ethics should also indicate the right of the accused and the accusers should not act as judge and jury (Knight, 1991).

The enforcement procedures provide a process for receiving, investigating and adjudicating allegations. Society members are held to the *Code,* as are persons holding a society-sanctioned credential. The objective is to engage in a process that is fair, responsible and confidential.

The procedures are designed to protect society members from unfair allegations, and guard society officers and staff from personal liability in cases alleging unethical conduct.

The procedures should reflect seven fundamental principles:

1. to be considered, an allegation of violation of the *Code* must be made in writing and signed by an individual,
2. the written complaint must be filed with the office of the president and CEO, or to the International Headquarters, for consideration by the Ethics Committee,
3. complaints can be made/alleged only against members or a person holding an society-sanctioned credential,
4. laws take precedence over ethical misconduct,
5. the role of society chapters in the ethics process is to educate members about ethical issues and standards, the *Code,* and the enforcement procedures in general,
6. chapters have no formal or informal role in the processing or adjudication of complaints, and
7. chapters should focus their action on ethical issues, standards and education.

The Ethics Committee carries out the enforcement process if there is an alleged violation of the Code that is supported by a formal complaint and sufficient documentation. If efforts to persuade the person to cease and desist have failed, the Committee may decide to hold a hearing to determine whether a violation of the Code has occurred and decide what disciplinary action, if any is appropriate.

Penalties for violation of the Code include:

1. a letter of reprimand,
2. censure and prohibition against holding Association and chapter office or participating in society activities for one year,
3. suspension of membership and prohibition against participation in society activities for a stated period, and
4. permanent revocation of membership, including recommendation to withdraw any society sanctioned credential.

Measurement of the success of a code of ethics falls into two categories, process and outcome. Examples of process measures for an ethics and compliance program might include, for example:

1. number of employees attending ethics and compliance training,
2. Twenty per cent of employees who have signed the company code of conduct,
3. number of calls or contacts to the ethics helpline, and
4. the quality or completeness of the information in the ethics and compliance case database.

Process measures are straight-forward and look at events and activities. They are necessary to ensure things happen as planned and provide feedback on the use of resources. When it comes to ethics and compliance programs measuring effectiveness, it may be difficult but it should not be impossible to implement a code of ethics in any organization and, even though the organization might be unique, the code should the measures need to fit the needs and requirements of the organization.

Measuring the effectiveness of ethics and compliance programs is a developing discipline and may appear a daunting task.

One manner of measuring the effectiveness of a code of ethics is to observe:

1. levels of observed misconduct,
2. reporting observed misconduct, and
3. response to any reprimands or punishment metered out to those found guilty of misconduct and/or unethical behavior.

One of the difficulties involved in investigating claims of unethical behaviors in order to meaningfully address the issue of scientific misconduct at the international level stems from the diversity of definitions and procedures among countries and institutions. Any definition, although deemed satisfactory when designed, might be considered to be too restrictive when reduced to practice. Thus, alternatives to defining misconduct include attempts to take inventory of all misbehavior that undermine science or engineering integrity or proposals for a code of conduct.

However, investigation of misconduct *must* proceed sand not be delayed on the basis of a (flimsy) excuse related to procedure. At the same time, the proceedings cannot take the form of a *kangaroo court* presided over by a *hanging judge*!

At different stages, scientific and engineering misconduct tends to be overlooked insofar as events are underreported or not reported at all (Nitsch et al., 2005), leaders of institutions and organizations are reluctant to launch investigations, and journal editors sometimes do not retract erroneous papers.

On the other hand, leaders of institutions have their share of responsibility. They are sometimes reluctant to conduct investigations, since misconduct is likely to jeopardize the name of their institution/organization. Investigations can also be costly in terms of time and money, and sometimes institutions are poorly equipped to conduct them. However, the failure to proceed with an investigation of reported misconduct will always appear to be paradoxical to any observer since such investigations could show how the institution/organization is able to reliably handle misconduct and punish the miscreants.

Journal editors, and reviews of submitted papers, also have a role in disclosing information about the integrity of the papers. When fraud is assessed, some journals do not publish any note nor retract the paper, sometimes because they require that all the authors agree with the retraction. Whatever, the case, some action must be taken (Relman, 1989); it is unfortunate that this is not always the case and a cover-up (advertent or inadvertent) may be the *modus operandi* of the university or organization involved.

Published retractions play a valuable role because publicity has a key role in cleaning retracted papers off the literature. In fact, the less public coverage a retracted paper receives, the more likely it to continue to be cited.

For example, a professor (while developing a theory) did not check the credibility of the experimental work of his students and coworkers was informed that the experimental work was not only flawed but incorrect; the coworkers and students had not performed any control experiments to collect and assess the base date. Nevertheless, the professor continued to cite this paper as evidence for his theory until the errors and omissions were it was pointed out in another paper (by other workers) that the credibility of his earlier work was non-existent. The professor even tried one more time to cite his earlier work as evidence for his theory but the journal editor had the wherewithal to use one of the two opposing workers as a reviewers. He pointed out using direct and straight-to-the-point comments and since then the professor appears to have ceased and desisted in his quest to use his earlier (flawed) work as evidence for his theory.

Obviously, this professor was not concerned about the culture of his research school and was promoting dishonesty to his coworkers and students. The university, of course, was apprised of the details but refused to act in any manner.

To build a culture of research integrity, proper training of current and future researchers is essential. Any researcher is less likely to *misbehave* not only when there are internal rules of conduct but also where the environment is such that research misconduct in any form is not overlooked – for any reason!

Integrity should be developed in an overall research education program, in a creative way by the professor, with support or guidance from the university. In the above example, this did not happen.

Investigation of claims of unethical behavior by a society member are usually forwarded to the society Ethics Committee, which is a committee dedicated to the rights and well-being of the society members; the Ethics Committee may also be known as a Review Board.

For example, the American Chemical Society has a Committee on Ethics whose mission is:

> To coordinate the ethics-related activities of the Society, serve as an educational resource and clearinghouse, but not as an adjudication body, for ACS members seeking guidance on ethics issues; raise awareness of ethics issues through meeting programming and columns/editorials; review recognition opportunities for acknowledging ethical behavior; and to develop and oversee such other ethics-related activities as will serve ACS members and promote the Society's standards of ethical conduct within the profession of chemistry and its related disciplines.

This Committee not only oversees ethical issues as they relate to the society but also maintains a vigil of ethics-related issues so that members are fully aware of ethical issues and can seek guidance on such issues.

In any society, the Ethics Committee is an advisory group appointed by the Executive Board. The Committee reviews, on request, ethical or moral questions that may arise from a member. Committee members include members of the society from all sub-disciplines.

Upon request, the committee will provide advisory consultation and review in cases where ethical dilemmas are perceived by any member.

The primary responsibility will be:

1. to encourage dialogue,
2. to identify issues offer viable options,
3. to seek supplemental resources,
4. to encourage problem resolution at the physician/patient level, and most important
5. to not pass the buck.

For example, the sole advisor (who was a non-academic) to a PhD student discovered that that student always managed to have an excuse for not attending a pre-arranged update meeting.

Upon reading the draft thesis, the advisor discovered that the student had been guilty of:

1. plagiarism,
2. included work in the thesis that was not his own but which he passed off as his own, and
3. unethical behavior in which he (the student) had enlisted the aid of others (academic faculty members) to justify not reporting the work to the his-academic advisor.

The advisor submitted a detailed report to the Provost and Vice-Provost who passed it to the Ethics Committee (composed of the Provost and Vice-Provost!) of that university. The Committee justified the student's actions seven-ways-from-Sunday and decided that the student was in the right – by inference, the non-academic advisor was unjustified in his claims of impropriety. The student was allowed to graduate with a PhD.

One wonders about the type of message this decision sent to other students and to other faculty when a non-academic advisor is seen (by inference) as the villain of the piece.

The expected penalty for engaging in an unethical action is the product of a penalty specified for doing so and a probability that the penalty will be borne. Organizations have two avenues when designing enforcement provisions: To increase the size of the penalty or to increase the probability that an individual engaging in an unethical action will bear that penalty. The greater the expected penalty, the less attractive an unethical action is to a decision maker, thus discouraging him or her from engaging in that action.

Although, in general, a very large penalty will have a greater impact on discouraging unethical actions than will a small penalty, organizations face significant constraints on the effective size of the penalty.

In designing an enforcement mechanism, organizations should recognize these constraints:

1. being subject to the code may be voluntary, and
2. economic alternatives available to the decision maker may allow them to escape sanctions by changing jobs or withdrawing from the organization.

Thus, an organization that is designing or modifying a code of ethics with the goal of having an impact on the action that a decision maker takes has two options: (1) educate the employees or membership as to the ethical nature of an action, and (2) create a disincentive which discourages the employees or membership from taking an unethical action. Both options apply to workability of the code of ethics as well as to the enforcement mechanisms.

The challenge for many higher education institutions is to cultivate a culture of inquiry that puts the emphasis on professional responsibility, obligations, commitments and research. Some class-based ethical issues are: using the work (assignments) given to students as data for research, balancing a power of coercion by a professor with sensitivity to the concerns of the students and prioritizing the educational value of a course over concerns about achieving research goals through teaching of the course. (Gale, 2002, page 40; Markle, 2002, page 42). In addition, there must also be a balance on the amount of research that is conducted by the professor and this should *not* affect preparation for teaching and to the point where students become discouraged and feel they are not learning (Braxton and Bayer, 1999, page 22–39 and page 43–60). This surely leaves the student (who might be readily tempted) no option but to seek alternate routes of submitting passing assignments and to seek the necessary means (other than intensive study) to pass examination.

Furthermore, the author learned the details of the revenue that the faculty generate for the university through a synopsis of a report prepared for the Texas A&M system (TAMU system). The report, which was not available to the authors of this book but was available as a synopsis in C&E News (C&EN, 2010b), listed salaries, money brought in from teaching, and research grant funding; the report was compiled in response to demands by Texas taxpayers for greater accountability from the university. The academics were not happy as it was not obvious to them how the report will influence their careers within the TAMU system. However, there is also concern that faculty could be pressured to focus on teaching, narrowly defined to emphasize classroom instruction, rather than the so-called broader aspects of scholarship, which includes research. Perhaps the thinking of non-academics has been off-track all of the years by thinking that the prime mandate of a university was to teach!

In summary, one of the best ways to prevent misconduct for any organization is to adopt and enforce policies which provide sufficient detail (written in understandable language to provide guidance to research staff (Taylor et al., 2006; Nadelson, 2007). If there are a wide variety of disciplines within the institution, then each discipline needs a data policy that is relevant to the specific type of research. In addition, the laboratory director should provide mentoring or formal training on policies (Hileman, 1997, 2005).

9.4 Reporting Misconduct

There is a growing consensus that researchers have a responsibility to report misconduct by other researchers when the misconduct is serious and when they are in a position to document it. Yet typically there are strong pressures (from supervisors, colleagues, and others) not to report misconduct, and hence most instances fall into the category of whistleblowing. Measures to protect individuals who responsibly report research misconduct are being implemented at research facilities, and the concept of research misconduct now applies to punitive measures taken against these individuals (Von Hipple and Chalk, 1979; Bok, 1980; Sprague, 1989). However, there often remains the stigma against a person who reports a colleague of misconduct (Martin and Schinzinger, 2005).

There has been increased attention paid in the last 30 years to whistleblowing, both in government and in private industry. According to the codes of ethics of the professional scientific and engineering societies, scientists and engineers are compelled to blow the whistle on acts or projects that harm these values. Scientists and engineers also have the professional right to disclose wrongdoing within their organizations and expect to see appropriate action taken.

A *whistleblower* is a person who raises a concern about wrongdoing occurring in an organization or body of people and usually this person would be from that same organization.

The revealed misconduct may be classified in many ways; for example, a violation of a law, rule, regulation and/or a direct threat to public interest, such as fraud, health/safety violations, and corruption. Whistleblowers may make their allegations internally (for example, to other people within the accused organization) or externally (to regulators, law enforcement agencies, to the media or to groups concerned with the issues).

Most whistleblowers are *internal whistleblowers*, who report misconduct to a fellow employee or superior within their company. Internal whistleblowing occurs when an employee goes over the head of an immediate supervisor to report a problem to a higher level of management. Or, all levels of management are bypassed, and the employee goes directly to the president of the company or the board of directors. However it is done, the whistleblowing is kept within the company or organization.

One of the most interesting questions with respect to internal whistleblowers is why and under what circumstances people will either act on the spot to stop illegal and otherwise unacceptable behavior or report it. There is some reason to believe that people are more likely to take action with respect to unacceptable behavior, within an organization, if there are complaint systems that offer not just options dictated by the organization, but a *choice* of options for individuals, including an option that offers near absolute confidentiality.

External whistleblowers, however, report misconduct to outside persons or entities. In these cases, depending on the information's severity and nature, whistleblowers may report the misconduct to lawyers, the media, law enforcement or watchdog agencies, or other local, state, or federal agencies.

Internal whistleblowing and external whistleblowing are generally perceived as disloyalty, putting the whistleblower on the defensive. However, keeping the report of misconduct within the company is often seen as less serious than going outside of the company. Under most U.S. federal whistleblower statutes, in order to be considered a whistleblower, the federal employee must have reason to believe his or her employer has violated some law, rule or regulation; testify or commence a legal proceeding on the legally protected matter; or refuse to violate the law.

Anonymous whistleblowing occurs when the employee who is reporting misconduct refuses to divulge his name when making accusations. These accusations might take the form of anonymous memos to upper management or in the form of anonymous phone calls to the police. The employee might also talk to the news media but refuse to let her name be used as the source of the allegations of wrongdoing. On the other hand, *acknowledged whistleblowing*, on the other hand, occurs when the employee puts his name behind the accusations (thereby leading to the subcategories of *internal whistleblowing* and *external whistleblowing*) and is willing to withstand the scrutiny brought on by his accusations.

In order to bring accusations of misconduct, the whistleblower must be in a very clear position to report on the problem – hearsay is not adequate and first-hand knowledge is essential to making an effective case about wrongdoing. This implies or indicates that that the whistleblower must have enough expertise in the area to make a true assessment of the perceived misconduct. If a scientist or engineer has undertaken work in areas outside his expertise, then he may not be in a position to make a realistic and believable assessment of the perceived misconduct.

It is important for the whistleblower to understand his motives before reporting the perceived misconduct, and it will be only perceived until proven. The whistleblower *must not* take the action as a means of revenge upon fellow employees or the organization. Furthermore, it is *not* acceptable to blow the whistle in the hopes of future gains, such as promotion, or any form of public recognition, or financial gain.

Whistleblowers frequently face reprisal – sometimes at the hands of the organization or group which they have accused, sometimes from related organizations, and sometimes under law.

In cases where whistleblowing on a specified topic is protected by statute, U.S. courts have generally held that such whistleblowers are protected from retaliation. However, a closely divided US Supreme Court decision held that the First Amendment, free speech guarantees for government employees do not protect disclosures made within the scope of the employees' duties.

Ideas about whistleblowing vary widely. Whistleblowers are commonly seen as selfless martyrs for public interest and organizational accountability. Others view them as a tattle tale or a *snitch* (street slang), solely pursuing personal glory and fame. Some consider that whistleblowers should at least be entitled to a rebuttable presumption that they are attempting to apply ethical principles in the face of obstacles and that whistleblowing would be more respected in governance systems if it had a firmer basis in virtue ethics.

It is probable that many people do not even consider blowing the whistle, not only because of fear of retaliation, but also because of fear of losing their relationships at work and outside work.

Because the majority of cases are very low-profile and receive little or no media attention and because whistle-blowers who do report significant misconduct are usually put in some form of danger or persecution, the idea of seeking fame and glory may be less commonly believed.

Persecution of whistleblowers has become a serious issue in many parts of the world. Although whistleblowers are often protected under law from employer retaliation, there have been many cases where punishment for whistleblowing has occurred, such as termination, suspension, demotion, wage garnishment, and/or mistreatment by other employees. For example, in the United States, most whistleblower protection laws provide for limited *make whole* remedies or damages for employment losses if whistleblower retaliation is proven. However, many whistleblowers report there exists a widespread "shoot the messenger" mentality by corporations or government agencies accused of misconduct, and in some cases whistleblowers have been subjected to criminal prosecution in reprisal for reporting wrongdoing.

As a reaction to this many private organizations have formed whistleblower *legal defense funds* or support groups to assist whistleblowers. Depending on the circumstances, it is not uncommon for whistleblowers to be ostracized by their co-workers, discriminated against by future potential employers, or even fired from their organization. This campaign directed at whistleblowers with the goal of eliminating them from the organization is referred to as *mobbing*. It is an extreme form of workplace bullying wherein the group is set against the targeted individual.

In the United States, legal protections vary according to the subject matter of the whistleblowing, and sometimes the state in which the case arises. Nevertheless, a wide variety of federal and state laws protect employees who call attention to violations, help with enforcement proceedings, or refuse to obey unlawful directions.

The collection of laws means that victims of retaliation need to be alert to the laws at issue to determine the deadlines and means for making proper complaints.

Some deadlines are as short as 10 days while it is 30 days for environmental whistleblowers to make a written complaint to the Occupational Safety and Health Administration (OSHA). Federal employees complaining of discrimination, retaliation or other violations of the civil rights laws have 45 days to make a written complaint to their agency's equal employment opportunity (EEO) officer. Airline workers and corporate fraud whistleblowers have 90 days to make their complaint to OSHA. Nuclear whistleblowers and truck drivers have 180 days to make complaints to OSHA. Victims of retaliation against union organizing and other concerted activities to improve working conditions have six months to make complaints to the National Labor Relations Board (NLRB). Private sector employees have either 180 or 300 days to make complaints to the federal Equal Employment Opportunity Commission (depending on whether their state has a deferral agency) for discrimination claims on the basis of race, gender, age, national origin or religion. Those who face retaliation for seeking minimum wages or overtime have either two or three years to file a civil lawsuit, depending on whether the court finds the violation was *willful*.

Those who report a false claim against the federal government, and suffer adverse employment actions as a result, may have up to six years (depending on state law) to file a civil suit for remedies under the US False Claims Act (FCA). Under this Act, the *original source* for the report may be entitled to a percentage of what the government recovers from the offenders. However, the *original source* must also be the first to file a federal civil complaint for recovery of the federal funds fraudulently obtained, and must avoid publicizing the claim of fraud until the US Department of Justice decides whether to prosecute the claim itself. Such lawsuits must be filed under seal, using special procedures to keep the claim from becoming public until the federal government makes its decision on direct prosecution.

Legal protection for whistleblowing varies from country to country but most countries have a framework of legal protection for individuals who disclose information so as to expose malpractice and matters of similar concern. In the vernacular, it protects whistleblowers from victimization and dismissal.

There are ways for an organization to solve the whistleblowing problem and theses include

1. a strong organizational ethics culture, which includes a clear commitment to ethical behavior,
2. clear lines of communication within the corporation, which gives openness to an employee who feels that there is something that must be fixed a clear path to air his concerns,
3. all employees must have meaningful access to high-level managers in order to bring their concerns forward, and
4. willingness on the part of management to admit mistakes, publicly if necessary (Martin and Schinzinger, 2005).

9.5 Published Examples of Unethical Behavior

There are many examples of unethical behavior or misconduct in science and engineering (Jackson, 1981; Hileman, 1997; Resnik, 1998; Hileman, 2005; Martin and Schinzinger, 2005; Fleddermann, 2008) – too many to be reproduced here – so only the most recent salient published examples will be presented.

Many scientist and engineers find that discovering unethical behavior among co-workers actually tests their own values and ethical principles. After all, unethical behavior that is not illegal frequently falls in a grey area between right and wrong that make it difficult to decide what to do when it is encountered. Furthermore, different scientist and engineers have different views regarding what is ethical and what is unethical. For example, some people feel that it is alright to tell a little *white lie*, or to make one long distance call on the company's time and money, as long as they can justify it in their mind.

A *white lie* is, by a misused and misinformed definition, an unimportant lie (especially one told to be tactful or polite). However, in truth, it is a lie (also called prevarication, falsehood) and is a type of deception in the form of an untruthful statement, especially with the intention to deceive others, often with the further intention to maintain a secret or reputation, protect someone's feelings or to avoid a punishment or repercussion. In short, it is a deliberate, untrue statement which (supposedly) does no harm or is intended to produce a favorable result for the originator of the white lie.

When scientist and engineers discover colleagues doing something that they know is wrong by the company's standards, their own sense of what is right and what is wrong instantly comes into question. That person needs to consider how he feels about that particular activity, as well as informing about that activity, or turning a blind eye. Even by deciding to do something about it, the scientist or engineer who has discovered the unethical behavior is presented with a number of difficult choices. Should he speak to the perpetrator directly? Should he arrange to report the incident directly to a company supervisor?

To make this decision easier, many companies have adopted several techniques that allow for the management of unethical activities. The first step is to create a company policy, in writing, that is read and signed by each employee. This erases most feelings of ambiguity when it comes to deciding what to do after witnessing an unethical behavior.

The second is to give a clear outline of what is expected of the person who has discovered the unethical behavior.

It should include the person who should be contacted, and how to go about doing it. With clear instructions, there will be less hesitation in reporting unethical activities, and then they can be dealt with quickly and relatively easily, before they develop into overwhelming issues.

Furthermore, the repercussions of unethical behavior should be clearly stated – such as summary dismissal – and acted upon when accusations of misconduct by scientist or engineer are proven (or admitted). This way, both the person doing the activity, and the witness to the activity will be well aware of the way that things will be dealt with, and there won't be any risk of someone not reporting unethical behavior because they're afraid that the culprit will be unfairly treated.

In a recent issue of Nature magazine there was an interesting report of an examination of misconduct (Titus et al., 2008). The article reported that a poll of 3,247 scientists, asking a range of questions relevant to scientific misconduct. The study showed that 1 in 3 scientists has been guilty of fudging their results. Of 3,247 early- and mid-career researchers who responded, less than 1.5% admitted to falsification or plagiarism, the most serious types of misconduct listed. But 15.5% said they had changed the design, methodology or results of a study in response to pressure from a funding source; 12.5% admitted overlooking others' use of flawed data; and 7.6% said they had circumvented minor aspects of requirements regarding the use of human subjects. Overall, about a third admitted to at least one of the ten most serious offences on the list, which is a range of misbehavior described by the authors as 'striking in its breadth and prevalence.

The fact that many scientists are willing to tell an intellectual white lie seems fitting, given the circumstances. Still, their actions are reprehensible.

An increasing number of scientists are allured by flashy results and quick publications that will lead to widespread publicity. Again, given current trends in science, the results of this study are disheartening and disappointing but certainly not surprising.

As examples of what are being claimed to be misconduct by scientist and engineers, several recent issues that have made headlines are given below. In order to identify these examples, the headlines used in the media are presented.

It is not the intent here to act as judge and jury but merely to report what had been found and presented elsewhere. The reader can then decide on the issues for himself whether or not he requires more detail of each case.

Giving Proper Credit (C&EN, 2007)
In 2007, after an extensive investigation in Stockholm University (Sweden) sanctioned an associate professor of chemistry, Armando Córdova, for research misconduct. In a number of cases, the investigation found that Córdova violated scientific ethics in his quest to publish research results in the emerging field of organo-catalysis. It appears that Córdova failed to cite or cite properly the work of other scientists and thereby taking credit for new discoveries that were not his own.

It was also reported that this case revealed that the scientific community is often unprepared to deal with misconduct, particularly when the violations fall short of scientific fraud. Although ethical guidelines themselves seem clear, what to do about ethics violations is another matter.

Whatever, the issues, universities must learn that punishment should match the severity of the violations.

Bell Labs Confirms Fraud (C&EN, 2002)
Suspicions that Hendrik Schön, formerly a researcher at Bell Laboratories, was responsible for falsifying and fabricating scientific data were recently confirmed. An independent

inquiry commissioned by Bell Labs concluded that a member of one of the company's research teams had engaged in scientific misconduct. A spokesman verified that the scientist had been identified as Schön and that his employment with Bell Labs was terminated.

A committee of scientists and engineers investigated the validity of research data in the areas of molecular electronics (Schön field of interest), superconductivity and molecular crystals. The committee concluded that data manipulation and data misrepresentation has occurred. All other researchers who had contributed to the work in question were cleared of any misconduct.

Scientist Guilty on 47 Counts (C&EN, 2003)
A jury in Lubbock, Texas, has found plague researcher Thomas C. Butler guilty of 47 of the 69 charges he originally faced. He was cleared of one of the most serious charges: that he lied to FBI agents on January 14 when he told them that 30 vials of plague bacterium were missing from his Texas Tech University laboratory. Among other charges, the jury convicted him of having defrauded Texas Tech via contracts he had with pharmaceutical companies as well as having shipped plague cultures to Tanzania without proper permits or labeling. Butler spent two years in prison but was later acquitted of smuggling plague.

Chemistry's Colossal Fraud (Chemistry World, 2008)
One of the biggest cases of scientific fraud in chemistry occurred in India and involves senior academics who co-authored a considerable number of discredited academic papers with researcher Pattium Chiranjeevi.

An enquiry committee last year found Chiranjeevi, a professor of chemistry at Sri Venkateswara University (SVU), Tirupati, guilty of plagiarizing or falsifying results in over 70 journal articles published between 2003 and 2007. The case saw the researcher stripped of all responsibilities except teaching – he continues to protest his innocence and is preparing to take legal action.

It is felt that the researchers/professors from other faculties that put their names to nearly 45 of the suspect papers include the heads of the university's physics, mathematics, geology and environmental sciences departments should have vetted the data before either putting their names (or allowing their names to be put) to a paper.

Climate Scientist Steps Down (WSJ, 2009)
The British scientist, Philip Jones, at the center of a scandal over climate-change research temporarily down as director of the Climatic Research Unit as a result of an internal probe that followed the release of hacked emails involving him and other scientist.

The action arose after hackers stole emails and documents from the East Anglia center that suggested Dr. Jones and other similar-thinking scientists tired to cover up the views of dissenting researchers and advocated manipulating the data. Dr. Jones defended the integrity of the institute's scientific work, while saying that he and his colleagues acknowledge that some of the published emails do not read well.

In addition, Pennsylvania State University confirmed that Michael Mann (see: *Judge Halts Virginia Climate Probe*), a climate scientist on the faculty who figures prominently in the emails, is under inquiry by the university. Dr. Mann's work reconstructing historic global temperatures has, over the past decade, become a focal point of debate.

As an aside and a point that seems to be forgotten (or ignored) in all of the climate-related debates and publications is that the earth is resilient to changes (Will, 2010) and also is currently in an inter-glacial period. As a result (surprise, surprise!) the temperature of the earth will increase. The actual extent of the temperature rise is unknown (who was around to measure the temperature increase during the last inter-glacial period?) but and will contribute to the overall temperature rise. Perhaps the scientists who ignore such a phenomenon are also guilty of misconduct.

Judge Halts Virginia Climate Probe (C&EN, 2010a)
The case stems from a fraud investigation of climate researcher Michael E. Mann, who worked at the University of Virginia from 1999 to 2005, and whether or not Mann committed fraud in connection with four federal grants and one state grant. Mann, who is currently the Director of Earth System Science Center at Pennsylvania State University, is among the groups of climate scientists whose controversial e-mails were hacked from the University of East Anglia, in England, and made public late in 2009. The University of Virginia fought to withhold release any of Mann's original documents and has been successful in court. However, it is not clear what Mann did that was misleading, false, or fraudulent in obtaining funds from the Commonwealth of Virginia for his research.

No doubt there will be more to this issue as time goes by.

Academic Fraud in China (Economist, 2010)
It would seem that fraud remains rampant in China and misconduct ranges from falsified data to untruths about degrees, cheating on tests, and extensive plagiarism. The most notable recent case focuses on Tang Jun, ironically author of a popular book *My Success Can Be Replicated*, who has was recently been accused of falsely claiming that he had a doctorate from the California Institute of Technology. He responded that his publisher had erred and in fact his degree is from another, much less prestigious, California school.

Other cases involve accusations of plagiarism against well-known Chinese scholars which have led to serious talk of investigations.

Such lapses of integrity do not appear to be unique to China, poor review mechanisms and a lack of checks on academic behavior all allow fraud to be more common. This calls into question the overall credibility of the scientific enterprise in China and leads to concerns about the safety of Chinese products and the integrity of information coming out of China.

Harvard Finds Scientist Guilty of Misconduct (NY Times, 2010)
In August, after a three-year internal investigation, Harvard University announced that it had found a prominent researcher, Marc Hauser, responsible for eight instances of scientific misconduct during the course of his work related to *cognition and morality*. Hauser has also done work suggesting that morality has an evolutionary basis in animals, and has written two well-received books on the evolution of cognition, morality, and behavior.

Hauser is, at the time of writing, was on a one-year leave of absence from Harvard and has acknowledged that he made some significant mistakes; he also apologized for the problems this case had caused to his students, his colleagues, and the university.

In response to the investigation's findings, the University's Dean of the Faculty of Arts and Sciences has vowed to determine the sanctions that are appropriate (http://arstechnica.com/science/news/2010/08/harvard-professor-found-guilty-of-scientific-misconduct.ars).

Michael Smith, the Dean at Harvard involved with this issue who wrote up the results of the investigation, acknowledges that Hauser was found guilty of scientific misconduct, but he also added that the university considers specific sanctions applied to anyone found responsible for scientific misconduct to be confidential the. It appears that Harvard will not officially release details of whatever sanction they deem appropriate.

References

Barber, T.X. 1976. *Pitfalls in Human Research: Ten Pivotal Points*. Pergamon Press, Toronto, Ontario, Canada.

Bok, S. 1980. "Whistleblowing and the Professional Responsibilities." *In Ethics Teaching in Higher Education*. D. Callahan and S. Bok (Editors). Plenum Press, New York. Page 277.

Braxton, J.M. and Bayer, A.E. 1999. *Faculty Misconduct in Collegiate Teaching*. John Hopkins University Press, Baltimore, Maryland.

Brown, T. 2007. "Ward Churchill's Twelve Excuses for Plagiarism." *Plagiary: Cross-Disciplinary Studies in Plagiarism, Fabrication, and Falsification*, 2(1): 1–11.

C&EN. 2002. "Fraud in the Physical Sciences." Reported by Mitch Jacoby. *Chemical & Engineering News*, May 27, page 17, and November 4, page 31.

C&EN. 2003. "Scientist Guilty on 47 Counts." Reported by William G. Schulz. *Chemical & Engineering News*, December 8, page 9.

C&EN. 2007. "Giving Proper Credit." Reported by William G. Schulz. *Chemical & Engineering News*, March 12, page 35.

C&EN. 2010a. "Judge Halts Virginia Climate Probe." Reported by Cheryl Hogue. *Chemical & Engineering News*, September 6, page 56.

C&EN. 2010b. "A Professor's Worth." Reported by Elizabeth Wilson. *Chemical & Engineering News*, September 20, page 4.

Carpenter, D.D., Harding, T.S., Finelli, C.J., and Passow, H.J. 2004. "Does Academic Dishonesty Relate to Unethical Behavior in Professional Practice? An Exploratory Study." *Science and Engineering Ethics*, 10(2): 311–324.

Chang, K. 2002. "On Scientific Fakery and the Systems to Catch It." *New York Times* (Late Edition, East Coast)), New York Times Company, New York, NY, October 15. Page F.1.

Chemistry World, 2008. *Chemistry's Colossal Fraud*. Reported by Killugudi Jayaraman. April, page 10.

Economist. 2010. "Academic Fraud in China: Replicating Success." *Economist*, 396(8692, July 24): 43.

Fleddermann, C.B. 2008. *Engineering Ethics 3rd Edition*. Pearson Prentice Hall, Upper Saddle River, New Jersey.

Frankel, M.S. 1989. "Professional codes: Why, how, and with what impact?" *Journal of Business Ethics*, 8(2–3):109–115.

Gale, R. 2002. "Commentary on Susan Burgoyne's Case." *In Ethics of Inquiry: Issues in the Scholarships of Teaching and Learning*. P. Hutchings (Editor). Carnegie Foundation for the Advancement of Teaching, Stanford, California. Page 39–40.

Glaser, B.G. 1964. *Organizational Scientists: Their Professional Careers*. Bobbs-Mettill, Indianapolis, Indiana.

Guba, E.G. 1990. "The Alternative Paradigm Dialog." In *The Paradigm Dialog*. E.G. Guba (Editor). Sage Publications, Thousand Oaks, California. Page 17–27.

Hileman, B. 1997. "Misconduct in Science Probed." *Chemical & Engineering News*, American Chemical Society, Washington, DC. June 23, page 24–25.

Hileman, B. 2005. "Research Misconduct." *Chemical & Engineering News*, American Chemical Society, Washington, DC. 83(35): 24–26.

Jackson, W. 1981. "Frauds in Science." *Reason and Revelation*, 1(2): 6–7.

Knight, J. 1991. "Scientific Misconduct: The Rights of the Accused." *In Issues in Science and Technology*. Fall 1991, page 28.

Lere, J.C. and B.R. Gaumnitz. 2003. "The impact of codes of ethics on decision making: Some insights from information economics." *Journal of Business Ethics*, 48(4):365–379.

Marcoux, H.E. 2002. "Kansas State University Faculty Perspective, Opinions, and Practices Concerning Undergraduate Student Academic Dishonesty and Moral Development." *A Dissertation Submitted in Partial Fulfillment of the Requirements for the Degree Doctor of Philosophy. Department Of Counseling And Educational Psychology College Of Education*, Kansas State University, Manhattan, Kansas.

Markle, P.J. 2002. "Commentary on Susan Burgoynes Case." In *Ethics of Inquiry: Issues in the Scholarships of Teaching and Learning*. P. Hutchings (Editor). Carnegie Foundation for the Advancement of Teaching, Stanford, California. Page 40–43.

Martin, M.W., and Schinzinger, R. 2005. *Ethics in Engineering 4th Edition*. McGraw Hill, New York.

McDowell, B. 2000. *Ethics and Excuses: The Crisis in Professional Responsibility*. Quorum Books, Westport, Connecticut.

Nadelson, S. 2007. "Academic Misconduct by University Students: Faculty perceptions and responses." *Plagiary: Cross Disciplinary Studies in Plagiarism, Fabrication, and Falsification*, 2(2): 1–10.

Nitsch, D., Baetz, M., and Hughes, J.C. 2005. "Why Code of Conduct Violations Go Unreported: A Conceptual Framework to Guide Intervention and Future Research." *Journal of Business Ethics*, 57(4): 327–341.

NY Times. 2010. "Harvard Research May Have Fabricated Data in Monkey Study." Reported by Nicholas Wade. *New York Times*, August 28, page A-11.

Relman, A.S. 1989. "Essay – Fraud in Science: Causes and Remedies." *Scientific American*, April, page 126.

Resnik, D.B. 1998. *Ethics in Science: An Introduction. Routledge*, New York.

Ryan, A. 2009. "Ethnography, Constitutive Practice and Research Ethics." *In The Handbook of Social Research Ethics*. D.M. Mertens and P.E. Ginsberg (Editors). Sage Publications, Thousand Oaks, California. Page 229–242.

Schwartz, M. 2001. "The Nature and Relationship between Corporate Codes of Ethics and Behavior." *Journal of Business Ethics*. 32(3):247–262.

Sprague, R. 1989. *A Case of Whistleblowing in Research. Perspectives on the Professions*, 8: 4.

Taylor, K., Paterson, B., Usick, B., Thordarson, J. & Smith, L. 2006. *Preventing Plagiarism: The Intersection of Education, Discipline, Policy and Practice*. Presented at AERA National Conference, San Francisco, California. April.

Titus, S.L., Wells, J.A., and Rhoades, L.J. 2008. Repairing Research Integrity. Nature 453(June 19): 980–982.

Von Hippel, F., and Chalk, R. 1979. "Due Process for Dissenting Whistle-blowers." *Technology Review*, June/July, page 49.

Werner, Bernard. 1995. Judgments *of Responsibility: A Foundation for A Theory of Social Conduct. New York:* The Guilford Press.

Will, G.F. 2010. "The Earth Doesn't Care About What is Done To or For it." *Newsweek*, September 20, page 26.

Woodward, J., and Goodstein, D. 1996. "Conduct, Misconduct, and the Structure of Science." *American Scientist*, 84: 481–490.

WSJ. 2009. "Climate Scientist Steps Down." *Wall Street Journal*, December 2, page A3.

Glossary

Absolutism: The belief that there is one and only one truth; those who espouse absolutism usually also believe that they know what this absolute truth is. In ethics, absolutism is usually contrasted to relativism.

Academic Freedom: The liberty or privilege that academics enjoy in regard to teaching, research and publications.

Academic honesty: Academic honesty or integrity is the maintenance of truthfulness and proper crediting of sources of ideas and expressions. Behaviors such as cheating on examinations and lab reports, or plagiarism of course papers and homework assignments, violate academic integrity. Other matters of academic integrity include honesty in writing letters of recommendation and in reporting institutional statistics.

Academic integrity: See Academic honesty.

Accountable: To be accountable is to be answerable or required to answer for one's actions. Used with a moral connotation ("normatively") meaning morally required to answer for one's actions without specifying to whom one is accountable. Also used descriptively to describe the sociological fact that a person or organization in question is required to answer to a particular party by some rules or organizational structure.

Aesthetic values: Non-moral values, such as beauty, which are based on personal perceptions or preferences.

Altruism: A selfless concern for other people purely for their own sake – usually contrasted with selfishness or egoism in ethics.

Applied ethics: The direct and technical application of 'expert' normative ethical theories and principles to guide moral problems in family, work, and community. The term now is often used pejoratively to indicate unscholarly and unreflective, almost ideological, prescriptive moral judgments, sometimes including an abdication of individual responsibility when making moral judgments.

Applied research: Applied research is the investigation of phenomena to discover whether their properties are appropriate to a particular need or want, usually a human need or want. In contrast, basic research investigates phenomena without reference to particular needs and wants. Applied research is more closely associated with technology, engineering, invention, and development. Basic research is sometimes described as "pure research."

Aspirational: A strong desire to achieve something high or great. An aspirational code would be intended to reach a higher ethics standard that supersedes being in compliance.

Attributes of external environment: The tightness of employment market for scientists and engineers, presence of high profile cases of scientific fraud or human subjects risks, political climate for science and engineering as reflected in such activities as legislative efforts to manage the conduct of science and engineering in spheres like data access, new federal rules on oversight of federal research grants, degree of public openness, and respect for science and engineering.

Autonomy: The ability to freely determine one's own course in life. Etymologically, it goes back to the Greek words for "self" and "law." This term is most strongly associated with Immanuel Kant, for whom it meant the ability to give the moral law to oneself.

Benchmarking: The process of comparing one's ethics climate to that of a previously established "best practices" climate.

Blurred Relations: These are relationships between persons, events or variables that are unclear or have no definitive boundaries.

Casuistry: Developing general principles of ethics from analysis of collections of existing real cases with clear and generally undisputed judgments. Analogous to case law in the legal system.

Characteristics of scientific societies: The general role of scientific society in representing or speaking on behalf of discipline, range and types of ethics activities, staff commitments or membership involvement in ethics-related work, pervasiveness of support for ethical guidelines through presence of code, adoption of code by academic departments or related disciplinary associations.

Common morality: Accepted and usually pluralistic norms of human conduct developed through traditions of a society, community, or specific social group.

Code of Conduct or Code of Ethics: A central guide and reference for users in support of day-to-day decision making. It is meant to clarify an organization's mission, values and principles, linking them with standards of professional conduct. As a reference, it can be used to locate relevant documents, services and other resources related to ethics within the organization.

Code of Conduct: A listing of required behaviors, the violation of which would result in disciplinary action. In practice, used interchangeably with Code of Ethics.

Code of Ethics: Often conveys organizational values, a commitment to standards, and communicates a set of ideals. In practice, used interchangeably with Code of Conduct. Such standards as are reasonably necessary to promote – (1) honest and ethical conduct, including the ethical handling of actual or apparent conflicts of interest between personal and professional relationships; (2) full, fair, accurate, timely, and understandable disclosure in the periodic reports required to be filed by the issuer; and (3) compliance with applicable governmental rules and regulations.

Code Provisions: The specific standards of behavior and performance expectations that your organization chooses to highlight and address in your code.

Conflict of Interest: A person has a conflict of interest when the person is in a position of trust which requires her to exercise judgment on behalf of others (people, institutions, etc.) and also has interests or obligations of the sort that might interfere with the exercise of her judgment, and which the person is morally required to either avoid or openly acknowledge.

Consequentialism: Any position in ethics which claims that the rightness or wrongness of actions depends on their consequences.

Contexts of research: Laboratory versus field studies, experimental versus observational designs, individual-investigator versus collaborative projects, high versus low stakes outcomes, studies that are local versus national versus international in scope, insufficient versus adequate resources to achieve the goals of the research.

Corruption: The abuse of public power for private benefit. Perversion or destruction of integrity in the discharge of public duties by bribery or favor or the use or existence of corrupt practices, especially in a state or public corporation.

Counter-Example: An example which claims to undermine or refute the principle or theory against which it is advanced.

Credo: Fundamental beliefs (or a set of beliefs) or guiding principles.

Cross-Situational Behavioral Stability: Consistency of any individual's behavior across several settings: home, workplace, church, community group.

Culture-Agency Dynamic: The changing features of the unfolding interaction between cultural influences and individual's efforts to overcome such cultural constraints.

Decolonizing Research: Studies that were and are designed to reduce the effects of a colonial legacy thus empowering such people to redefine themselves.

Deductive Theorizing: Explanations that move from general to specific.

Deontology: The science related to duty or moral obligation. In moral philosophy, deontology is the view that morality either forbids or permits actions. For example, a deontological moral theory might hold that lying is wrong, even if it produces good consequences. Deductive. A deductive argument is an argument whose conclusion follows necessarily from its premises. This contrasts to various kinds of inductive arguments, which offer only a degree of probability to support their conclusion.

Developmental Research: Constitutive of studies that are intended to inform development activities practices.

Disciplinary culture: Modes of work typically through team or individual projects, normative practices and procedures that encourage consultation or make work processes as well as products transparent, amount of oversight due to presence of external funding or human subjects review, attention to ethics in curriculum.

Dispositional Factors: Qualities that are rooted in and individual's personality e.g. greed, honesty, composure.

Dispositional Orientations: Engagement in actions that clearly reflect features of one's personality e.g. honesty, friendliness.

Ego: That aspect of persons that orient them to think of themselves only.

Emotionality: The range of feelings experienced by an individual.

Emotivism: A philosophical theory which holds that moral judgments are simply expressions of positive or negative feelings.

Empathy: Caring about the consequences of one's choices as they affect others. Being concerned with the effect one's decisions have on those who have no say in the decision itself.

Enlightenment: An intellectual movement in modern Europe from the sixteenth until the eighteenth centuries that believed in the power of human reason to understand the world and to guide human conduct.

Entitlements: Whatever one has the right to receive.

Epistemological issues: Concerns about the nature of knowing; i.e. what does it mean to know? How do we know that we know?

Ethical currents: Invisible vibrations that communicate people's positions on ethical issues to all.

Ethical decision-making: Consideration of the right thing to do – as defined by the values and principles, which apply to a particular situation.

Ethical differences: Situations in which two people agree on a particular value and disagree as to the action to be taken or decision to be made.

Ethical dilemmas: Situations that require ethical judgment calls. Often, there is more than one right answer and no win-win solution in which we get everything we want.

Ethical egoism: A moral theory that, in its most common version (*universal ethical egoism*) states that each person ought to act in his or her own Self-interest. Also see Psychological Egoism.

Ethical values: Desired good – about what is right and wrong, what should or ought to be.

Ethics: The decisions, choices, and actions (behaviors) we make that reflect and enact our values. A set of standards of conduct that guide decisions and actions based on duties derived from core values. The morality of our actions, study of ethical theory and the philosophical reflection on morality's nature and function. Also, a set of moral principles resulting from reflection on morality. Synonyms: ethical theory, moral philosophy, moral theology, or philosophical ethics.

Ethnicity: A person's ethnicity refers to that individual's affiliation with a particular cultural tradition that may be national (such as French) or regional (such as Sicilian) in character. Ethnicity differs from race in that ethnicity is a sociological concept whereas race is a biological phenomenon.

Ethnographers: Researchers who utilize qualitative approaches observations, stories etc. to obtain information from several people on the topic that is being researched.

Ethos of Science: The culture or atmosphere that surrounds a particular discipline, in this case, science.

Functional ethics: Morality of complying to rules, codes, and laws.

Gender: A person's gender refers to that individual's affiliation with either male or female social roles. Gender differs from sex in the same way that ethnicity differs from race: gender is a sociological concept, while sex is a biological one.

Good faith: Based on the belief in the accuracy of the information or concern being reported.

Gray areas: Situations in which the individual's business standards lack clarity. The lack of clarity may be due to an individual's not being familiar with a guideline or a guideline that is vague and subject to interpretation. Guidelines are often written to provide managers with as much latitude as appropriate, and this may create gray areas.

Higher Education: The realm of education that should take an individual's level of development, in terms of knowledge, attitude and skills, to a more refined or higher level.

Impartiality: In ethics, an impartial standpoint is one which treats everyone as equal. For many philosophers, impartiality is an essential component of the moral point of view.

Individual attributes: History of professional integrity or lack thereof; exposure to explicit training in ethics or research conduct, track record of haste or careful planning in performing work, magnitude of professional or personal commitments, perception of the importance of a given publication for personal or group advantage.

In-house reporting system: Any system established by an organization to meet the standards of an effective program to prevent and detect violations of law in order to provide employees and other agents with a means to report misconduct to the organization without fear of retribution.

Integrationist: Any position which attempts to reconcile apparently conflicting tendencies or values into a single framework. Integrationist positions are contrasted with separatist positions, which advocate keeping groups (usually defined by race, ethnicity, or gender) separate from one another.

Integrity: Making choices that are consistent with each other and with the stated and operative values one espouses. Striving for ethical congruence in one's decisions.

Intellectual Property: This is the product of the intellect of any individual which may be in the form of a book, journal, poem or any other work.

Intersubjectivity: Interactions between the subjective aspects of the lives of people which happens continuously.

Lacunae: Shortcomings or deficiencies.

Maxims: Short, pithy statements that are used to instruct and guide behavior.

Moral Architecture: The layout and connections between various moral views in any specific place and time.

Moral Autonomy: Freedom to engage in behaviors irrespective of their rightness or wrongness of such outcomes.

Moral ballpark: The domain of actions, motives, traits, etc. that are open to moral assessment, that is, can be said to be morally good *or* morally bad.

Moral dilemmas: Dilemmas that occur when a person or an organization faces a decision with two or more compelling choices (to act or not to act or two or more courses of moral action) and the right choice is not clear. Moral dilemmas involve decisions with two or more equally compelling moral obligations. Decisions involving moral dilemmas, issues of character, and competing ways of thinking are critical judgments.

Moral isolationism: The view that we ought not to be morally concerned with, or involved with, people outside of our own immediate group. Moral isolationism is often a consequence of some versions of moral relativism.

Morality: Traditional, social, theoretical, or other conventions or belief about what is right and wrong, what should or ought to be. Does not capture the sense of moral reflection embodied in the term ethics.

Moral luck: The phenomenon that the moral goodness or badness of some of our actions depends simply on chance. For example, the drunk driver may safely reach home without injuring anyone at all, or might accidentally kill several children that run out into the street while the drunken person is driving home. How bad the action of driving while drunk is in that case depends in part on luck.

Morals: Values that are attributed to a system of beliefs that help the individual define right versus wrong, good versus bad. These typically get their authority from something outside the individual – a higher being or higher authority (e.g. government, society). Moral concepts, judgments and practices may vary from one society to another.

Moral Standards: Measures of decency or good thought or conduct which can be used to evaluate an individual's past, present and future behavior.

Mean: The arithmetical average of items in a group.

Natural law: In ethics, believers in natural law hold (a) that there is a natural order to the human world, (b) that this natural order is good, and (c) that people therefore ought not to violate that order.

Newcomer Socialization: Exposure of new appointees to the rules, regulations and ways of 'doing things' in any specific setting e.g. workplace.

Norms: Behavior patterns that facilitate preventing or detecting research misconduct (e.g., sharing of research data, replication studies, specification of research procedures, record keeping requirements such as documentation of interviews or detailed laboratory books, journal publication policies regarding review).

Objectivity-Subjectivity: Struggles within the self between efforts use reason or use feelings.

Ombudsman: A designated neutral or impartial dispute resolution practitioner whose major function is to provide confidential and informal assistance to managers and employees and/or clients of the employer: patients, students, suppliers or customers.

Phenomenological reduction: Interpreting all experiences in terms of their meanings.

Pluralism: The belief that there are multiple perspectives on an issue, each of which contains part of the truth but none of which contain the whole truth. In ethics, moral pluralism is the belief that different moral theories each capture part of truth of the moral life, but none of those theories has the entire answer.

Practical ethics: The reflection on morality by people in their everyday vocations of work, family and community.

Prima facie: In the original Latin, this phrase means "at first glance." In ethics, it usually occurs in discussions of duties. A *prima facie* duty is one which appears binding but which may, upon closer inspection, turn out to be overridden by other. stronger duties.

Professor: One who is an authority and provides leadership for the further development of a discipline, department or faculty.

Professional: An individual who prioritizes quality service to customers and all that would enhance same paying attention to the highest ethical standards.

Public ethics: Ethics associated with the public arena – public policy, institutional, external organizational ethics, and parts of professional ethics.

Rationality: Willingness and ability to reason.

Reciprocal Relationships: A mutual sharing and caring between individuals.

Relativism: In ethics, there are two main type of relativism. Descriptive ethical relativism simply claims as a matter of fact that different people have different moral beliefs, but it takes no stand on whether those beliefs are valid or not. Normative ethical relativism claims that each culture's (or group's) beliefs are right within that culture, and that it is impossible to validly judge another culture's values from the outside.

Religious ethics: Moral theology and theological ethics; reflects ethics based in religion or a religious tradition.

Research misconduct: A is a term used rather narrowly. It does not include all violations of standards of research ethics. In particular, it is not applied to violations of the norms for the use of human or animal subjects. In the United States the three actions that have been the focus of misconduct definitions are fabrication, falsification, and plagiarism. In 1995 the Congressionally mandated Commission on Research Integrity issued a report, "Integrity and Misconduct in Research," arguing that FFP did not cover all serious deviations from accepted practices, and proposed a broader definition of research misconduct as misappropriation, interference, and misrepresentation, but this definition was not adopted. After extensive public debate the U.S. Office of Science and Technology Policy in 2000 issued the following common definition: "Research misconduct is defined as fabrication, falsification, or plagiarism in proposing, performing, or reviewing research, or in reporting research results."

Rights: Rights are entitlements to do something without interference from other people (negative rights) or entitlements that obligate others to do something positive to assist you (positive rights). Some rights (natural rights, human rights) belong to everyone by nature or simply by virtue of being human; some rights (legal rights) belong to people by virtue of their membership in a particular political state; other rights (moral rights) are based in acceptance of a particular moral theory. Rules-centered code of conduct: Frequently takes the form of a list of behavioral requirements, the violation of which could result in disciplinary action.

Self as an object: That state when individuals are acted upon continuously, that is, external factors consistently shape and influence them.

Self as a subject: A phrase used to describe that state when individuals initiate actions and may make things happen.

Skepticism: There are two senses of this term. In ancient Greece, the skeptics were inquirers who were dedicated to the investigation of concrete experience and wary of theories that might cloud or confuse that experience. In modern times, skeptics have been wary of the trustworthiness of sense experience. Thus classical skepticism was skeptical primarily about theories, while modern skepticism is skeptical primarily about experience.

Sources of standards of conduct (e.g., promulgation of codes of ethics, mentors who have integrity, coursework that stresses good methodology and integrity).

Subjectivism: An extreme version of relativism, which maintains that each person's beliefs are relative to that person alone and cannot be judged from the outside by any other person.

Subjectivity: That which is reflected in situations where individuals allow their personal views, perceptions or feelings to influence their thoughts, interactions, research.

Subjectivity-to-objectivity: The shift from emotion-driven actions to rationality-driven behavior.

Transparency: Sharing information and acting in an open manner. A principle that allows those affected by administrative decisions, business transactions or charitable work to know not only the basic facts and figures but also the mechanisms and processes. It is the duty of civil servants, managers and trustees to act visibly, predictably and understandably.

Utilitarianism: A moral theory that says that what is morally right is whatever produces the greatest overall amount of pleasure (hedonistic utilitarianism) or happiness (eudaimonistic utilitarianism). Some utilitarians (act utilitarians) claim that we should weigh the consequences of each individual action, while others (rule utilitarians) maintain that we should look at the consequences of adopting particular rules of conduct.

Value freedom: Decisions that are not influenced by the values and beliefs of individuals.

Value judgments: Decisions that were or are influenced by an individual's values and beliefs.

Value neutrality: Used to describe a situation where individuals are neither for or against something.

Values: The core beliefs we hold regarding what is right and fair in terms of our actions and our interactions with others. Individual or shared conceptions of the desirable, goals considered worth pursuing.

Whistle-blower: A person who takes a concern (such as a concern about safety, financial fraud, or mistreatment) outside of the organization in which the abuse or suspected abuse is occurring and with which the whistle-blower is affiliated.

Whistleblowing: Informing on unethical behavior. Whistleblowing is made up of four components: (1) An individual act with the intention of making information public; (2) the information is conveyed to parties outside the organization who make it public and a part of the public record; (3) the information has to do with possible or actual nontrivial wrongdoing in an organization; (4) the person exposing the agency is not a journalist or ordinary citizen, but a member or former member of the organization.

Index

Also of Interest

Check out these other related titles from Scrivener Publishing

The Two Narratives of Political Economy, by Nicholas Capaldi and Gordon Lloyd, ISBN 9780470948293. Captures in one volume the 17th-19th century origins and developments of political economy by editing original texts and illuminating their relevance for today's political debate.

Global Bioethics, Edited by H. Tristram Engelhardt, Jr., ISBN 9780976404132. Twelve essays from international top class scholars analyze the issue of the repeated failure to derive a universal set of standards for bioethics and diagnose why consensus has been elusive.

The Greening of Petroleum Operations, by Rafiqul Islam, ISBN 9780470625903. This state-of-the-art text covers some of the most hot-button issues in the energy industry, covering new, greener processes for engineers, scientists, and students to move petroleum operations closer to sustainability.

The Two Faces of Liberalism, Edited by Gordon Lloyd, ISBN 9780976404125. Contains 60 original documents with notes that allows the reader to folow the dynamics of the debate between Hoover and FDR.

Emergency Response Management for Offshore Oil Spills, by Nicholas P. Cheremisinoff, PhD, and Anton Davletshin, ISBN 9780470927120. The first book to examine the Deepwater Horizon disaster and offer processes for safety and environmental protection.

Energy Storage: A New Approach, by Ralph Zito, ISBN 9780470625910. Exploring the potential of reversible concentrations cells, the author of this groundbreaking volume reveals new technologies to solve the global crisis of energy storage.

Zero-Waste Engineering, by Rafiqul Islam, February 2011, ISBN 9780470626047. In this controvercial new volume, the author explores the question of zero-waste engineering and how it can be done, efficiently and profitably.

Printed and bound by CPI Group (UK) Ltd, Croydon, CR0 4YY

16/04/2025

14658453-0001